Fourth International Conference: Shape Memory Alloys (SMA 2021)

IOP Conference Series: Materials Science and Engineering
Volume 1213

Moscow, Russia
13 - 17 September 2021

Editors:

Sergey Prokoshkin
Natalia Resnina
Sergey Dubinskiy
Yulia Zhukova
Vadim Sheremetyev
Victor Komarov
Kristina Polyakova

ISBN: 978-1-7138-4672-7
ISSN: 1757-8981

Printed from e-media with permission by:

Curran Associates, Inc.
57 Morehouse Lane
Red Hook, NY 12571

Some format issues inherent in the e-media version may also appear in this print version.

Institute of Physics
Dirac House, Temple Back
Bristol BS1 6BE UK

Phone: 44 1 17 929 7481
Fax: 44 1 17 920 0979

techtracking@iop.org

Additional copies of this publication are available from:

Curran Associates, Inc.
57 Morehouse Lane
Red Hook, NY 12571 USA
Phone: 845-758-0400
Fax: 845-758-2633
Email: curran@proceedings.com
Web: www.proceedings.com

TABLE OF CONTENTS

IOP Conf. Series: Materials Science and Engineering 1213 (2022) 011001 doi:10.1088/1757-899X/1213/1/011001

The Fourth International Conference "Shape Memory Alloys" (SMA 2021)

IV International Conference
SHAPE MEMORY ALLOYS
September 13-17, Moscow, Russia

The Fourth international conference "Shape Memory Alloys" (SMA 2021) continues the tradition of regular scientific events on shape memory alloys held in various cities of the Soviet Union: Kiev (1980, 1991), Voronezh (1982), Tomsk (1985), Novgorod (1989), Kosov (1991), St. Petersburg (1995). These days, the First conference "Shape Memory Alloys: properties, technologies, and prospects" was held in 2014 in Vitebsk (Belarus), the Second one – in 2016 in St. Petersburg, the Third one – in 2018 in Chelyabinsk.

The aim of the conference is to review modern research and development directions in the field of shape memory alloys and related phenomena: from studying their structure, physical, mechanical, and functional properties to mathematical modeling of the shape memory materials' behavior and their applications. The conference schedule comprised oral and poster presentations in the framework of three parallel sections:

- Structure, martensitic transformations and shape memory effects in alloys.
- The theory of martensitic transformations and shape memory effect: modeling and calculations.
- Novel materials. The manufacturing technology and application of shape memory alloys.

Editors of Proceedings: Sergey Prokoshkin, Natalia Resnina, Sergey Dubinskiy, Yulia Zhukova, Vadim Sheremetyev, Victor Komarov, Kristina Polyakova

Organizers:
National University of Science and Technology "MISIS"
Saint-Petersburg State University
"Interstate Coordinating Council for Physics of Materials Strength and Plasticity"

IOP Conf. Series: Materials Science and Engineering 1213 (2022) 011001 doi:10.1088/1757-899X/1213/1/011001

Organising Committee
The Chairman:
M.R. Filonov, Moscow, Russia
Co-chairs:
S.D. Prokoshkin, Moscow, Russia
N.N.Resnina, Saint-Petersburg, Russia
A.M. Glezer, Moscow, Russia

Scientific Secretary:
Dubinskiy, S.M, Moscow, Russia

Members of the organizing committee:
S.P. Belyaev, Saint-Petersburg, Russia
V.I. Betekhtin, Saint-Petersburg, Russia
V. Brailovski, Montreal, Canada
V.D. Buchelnikov, Chelyabinsk, Russia
R.Z. Valiev, Saint-Petersburg, Russia
A.E. Volkov, Saint-Petersburg, Russia
D.E.Gusev, Moscow, Russia
V.E. Gunther, Tomsk, Russia
S.V. Dobatkin, Moscow, Russia
Yu.S. Zhukova, Moscow, Russia
M.P. Kashchenko, Ekaterinburg, Russia
Yu.N. Koval, Kiev, Ukraine
A.S. Konopatsky, Moscow, Russia
A.V. Korotitskiy, Moscow, Russia
S. Kustov, Palma, Spain
A.I. Lotkov, Tomsk, Russia
A.A. Movchan, Moscow, Russia
R.R. Mulukov, Ufa, Russia
V.A. Plotnikov, Barnaul, Russia
N.N. Popov, Sarov, Russia
V.V. Rubanik, Vitebsk, Belarus
G.S. Firstov, Kiev, Ukraine
Yu.I. Chumlyakov, Tomsk, Russia

Program committee
The Chairman:
V.V. Khovaylo, Moscow, Russia

Program committee members:
V.A. Andreev, Moscow, Russia
I.V. Bychkov, Chelyabinsk, Russia
D.V. Gunderov, Ufa, Russia
V.I. Zel'Dovich, Ekaterinburg, Russia
L.M. Kaputkina, Moscow, Russia
I.V. Kireeva, Tomsk, Russia
V.V. Koledov, Moscow, Russia
M.Yu. Kollerov, Moscow, Russia
Yu.R. Kolobov, Chernogolovka, Russia
G.V. Markova, Tula, Russia
G.A. Malygin, Saint-Petersburg, Russia
E.Yu. Panchenko, Tomsk, Russia
M.I. Petrzhik, Moscow, Russia
V.G. Pushin, Ekaterinburg, Russia
A.I. Razov, Saint-Petersburg, Russia
V.V. Rubanik (jr), Vitebsk, Belarus
E.P. Ryklina, Moscow, Russia
V.V. Stolyarov, Moscow, Russia
I.Yu. Khmelevskaya, Moscow, Russia
V.G. Shavrov, Moscow, Russia
A.V. Shelyakov, Moscow, Russia
V.A. Sheremetyev, Moscow, Russia

Local Organizing Committee
Co-Chairs:
K.A. Polyakova, Moscow, Russia
V.S. Komarov, Moscow, Russia

Members of the local organizing committee:
R.D. Karelin, Moscow, Russia
A.A. Kudryashova, Moscow, Russia
P.O. Kadirov, Moscow, Russia
K.E. Lukashevich, Moscow, Russia
T.O. Teplyakova, Moscow, Russia
M.S. Tsaturyants, Moscow, Russia

IOP Conf. Series: Materials Science and Engineering 1213 (2022) 011002 doi:10.1088/1757-899X/1213/1/011002

Peer review declaration

All papers published in this volume of *IOP Conference Series: Materials Science and Engineering* have been peer reviewed through processes administered by the Editors. Reviews were conducted by expert referees to the professional and scientific standards expected of a proceedings journal published by IOP Publishing.

- **Type of peer review:** Single-blind

- **Conference submission management system:**

All the manuscripts received form the authors were sent to the reviewers. Each manuscript was reviewed by one reviewer and re-checked by editor. All comments were sent back to the authors to let them make all corrections. The revised versions of manuscripts were checked and approved by editors.

- **Number of submissions received:** 15

- **Number of submissions sent for review:** 15

- **Number of submissions accepted:**11

- **Acceptance Rate (Number of Submissions Accepted / Number of Submissions Received X 100):**73.3

- **Average number of reviews per paper:** 1

- **Total number of reviewers involved:**10

- **Any additional info on review process:**

Reviewers considered manuscripts in accordance with:

1) Technical Criteria: scientific rigour, accuracy, correctness of selected methodology.
2) Quality Criteria: originality and novelty, clarity of motivation and results importance.
3) Presentation Criteria: clarity of expression, readability and completeness of presentation, quality of all presented data and figures.

- **Contact person for queries:**

Name: Sergey Dubinskiy

Email: sdubinskiy@gmail.com

IOP Conf. Series: Materials Science and Engineering 1213 (2022) 012001 doi:10.1088/1757-899X/1213/1/012001

Functional properties of NiTi nanofilm/Kapton composite

A V Sibirev[1], M V Alchibaev[1], I A Palani[2], S Jayachandran[2], A Sahu[2], S P Belyaev[1] and N N Resnina[1]

[1] Saint Petersburg State University, 7/9 Universitetskaya nab., St Petersburg, 199034, Russia
[2] Department of Mechanical Engineering, Indian Institute of Technology Indore, Indore, 453552, India

Corresponding author: alekspb@list.ru

Abstract. The aim of the paper was to study the one-way and two-way shape memory effects in the NiTi nanofilm/Kapton composite. 500 nm film of the $Ni_{50}Ti_{50}$ alloy was deposited to Kapton by flash evaporation. After deposition, the NiTi layer was amorphous and the sample was held at a temperature of 350 - 400 °C for two hours in vacuum to crystallize the NiTi layer. As deposited sample as well as samples after heat treatment were bent around the mandrel with various diameters at room temperature and subjected to heating – cooling – heating through a temperature range of the martensitic transformations. It was shown that as-deposited sample did not demonstrate the recoverable stain variation. At the same time, the heat treated sample demonstrated the one-way shape memory effect on heating and a maximum recoverable strain was found to be 2 %. The two-way shape memory effect was not observed on further cooling and heating.

1. Introduction

Shape memory alloys (SMAs) are able to accumulate large unelastic strain on cooling and then recover it on heating due to the thermoelastic martensitic transformation. This unique property of SMAs allow them to be used as working element for various drives, sensors and actuators [1–4]. These devices can be produced in any shape or form, moreover they are easily scalable compared to traditional actuators (hydraulics, pneumatics or electrical motors), because the force is linearly dependent on effective cross-section area of SMA working body. The advantage of SMA actuators over traditional ones are higher generated forces, larger strain recovery, ease of production, smooth and reliable actuation [3,5–8]. Nowadays, using the SMAs actuators in microelectromechanical systems (MEMS) becomes most promising field [9–11]. To minimalize such SMA drives, different techniques were developed, and one of them was the film deposition on a flexible substrate. To produce thin films usually sputtering [8,12], thermal or flash evaporation [6,7,13] techniques are used, and the flash evaporation provides better quality nanofilms with uniform NiTi layers [13].

NiTi-based alloys are widely used for thin films production due to their superior functional and mechanical properties. There were many successful attempts to produce such films, however the thickness of the TiNi layer were more than several microns as it was difficult to produce uniform thin layer with crystalline structure [7,14]. Previously, the thin NiTi films with a thickness of 100 - 500 nm were deposited on Kapton by flash evaporation [13, 15]. The recoverable strain variation was studied on cooling and heating in stress free state or under a stress [15]. At the same time, the one-way and two-ways shape memory effects have never been studied in NiTi nanofilms deposited to Kapton that may be used in mircoactuators. Thus, the aim of the present work is to study the recoverable strain variation on heating – cooling – heating of the NiTi nanofilm/Kapton sample pre-deformed by bending around the mandrel with various diameters to find the influence of the residual strain on the one-way and two-way shape memory effects.

2. Materials and methods

The pieces of $Ni_{50}Ti_{50}$ wire with a total weight of 1 g were placed in tungsten boat (crucible) inside the high vacuum chamber. The vacuum inside the chamber was maintained at 5×10^{-5} mbar with rotary and diffusion pump during the flash evaporation deposition process. A $50 \times 20 \times 0.09$ mm^3 kapton polymide flexible sheet was used as the substrate for the NiTi nanofilm deposition. The polymide sheet was cleaned with isopropyl alcohol and deionized water before deposition and attached to substrate holder in flat position inside the chamber [15]. After deposition, the thickness of NiTi layer was measured by scanning electron microscopy and was equal to 500 ± 40 nm. X-ray analysis showed that the NiTi film was amorphous hence, to obtain the crystalline samples, the samples were held for 2 hours at 350 or 400 °C in vacuum (10^{-4} bar). The holding temperature could not be increased over 400 °C because the Kapton thermal dissociation occurs at 425 °C [16]. The structure of the films before and after heat treatment was studied by X-ray diffraction (XRD) analysis using a Bruker D8 DISCOVER diffractometer (CuK_{α}).

To characterize the functional properties, samples were preliminary deformed by bending around the mandrel with a diameter of 2, 1.2, 0.5 or 0.2 mm. During bending, the NiTi layer was on the outer side of the sample and do not touch the mandrel. To calculate the strain after unloading, the photos of the bent samples were digitized with OriginPro™ software as shown in fig. 1 and the bending radius was measured. The strain was calculated as $\varepsilon = \frac{h}{2r} * 100\%$, where h was sample thickness and r was a bending radius. After preliminary deformation, samples were installed in a holder and subjected to heating – cooling - heating in thermal chamber with a rate of 7 °C/min in the temperature range of 120÷0 °C. During temperature variation, the shape of the samples was photographed to estimate the strain variation and the sample temperature was measured with thermocouple attached to the sample. Thus, strain vs temperature dependences were obtained on heating – cooling – heating of the samples pre-deformed to various strain.

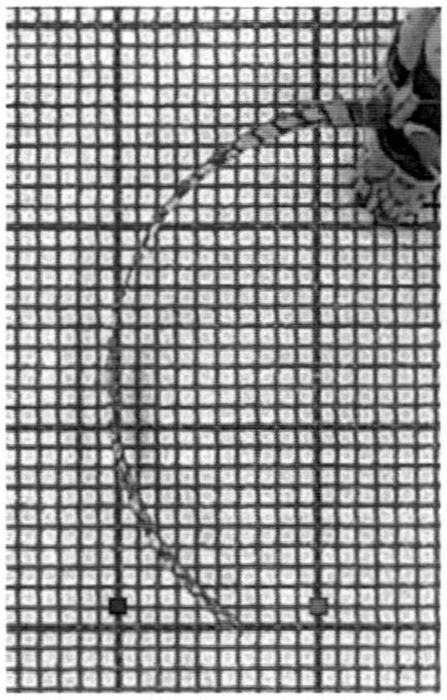

Figure 1. The photo and the digitizing the shape of the NiTi/Kapton sample pre-deformed around the mandrel.

IOP Conf. Series: Materials Science and Engineering 1213 (2022) 012001 doi:10.1088/1757-899X/1213/1/012001

3. Results and Discussion

Fig. 2 shows that the NiTi layer deposited on the Kapton was amorphous. Holding the sample at 400 °C for 2 hours led to the formation of thin oxide layer on NiTi side. The oxidation might be caused by oxygen escaped from Kapton due to holding temperature was close to dissociation temperature of Kapton (425 °C). The oxidation of the NiTi films decreases the volume of the NiTi film which is able to undergo the martensitic transformations, that worsens the functional properties. The oxidation of the NiTi surface was not observed in the sample held at 350 °C for 2 hours. Moreover, a martensite reflexes were found on the X-ray pattern of this sample (Fig. 2) that showed that thin NiTi layer underwent the crystallization during holding at 350 °C for 2 hours.

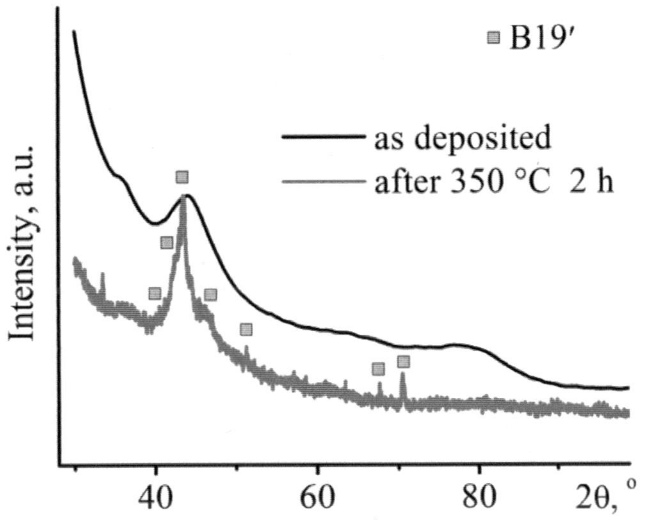

Figure 2. X-ray patterns obtained before and after heat treatment at 350 °C for 2 hours.

As deposited sample as well as the sample held at 350 °C for 2 hours were deformed at room temperature by bending around of the mandrels with different diameter and unloaded. The residual strain was measured in the samples after unloading and shown in Table 1. Then, samples were subjected to heating – cooling – heating in a temperature range of 120÷0 °C. It was found that as-deposited sample did not demonstrate any strain variation. At the same time, the strain recovery was found on heating of the heat treated sample (Fig. 3) hence, the NiTi/Kapton sample demonstrated the one-way shape memory effect. On subsequent cooling and heating, the strain variation was not found, thus, heat treated samples did not reveal the two-way shape memory effect.

Table 1. Radius of mandrel used for preliminary deformation and residual strain measured after deformation in each sample

N	R_m, mm	ε_{res}, %
1	1	0.8
2	0.6	0.1
3	0.25	1.1
4	0.1	2.8

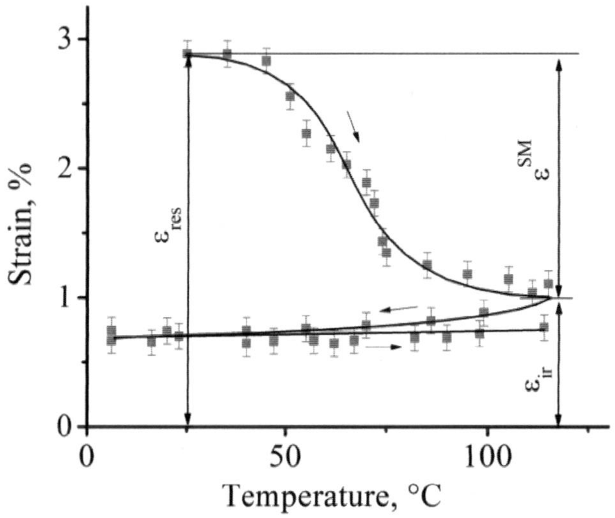

Figure 3. Strain on temperature variation in the NiTi/Kapton sample with a residual strain of 2.8%.

The shape memory effect (ε^{SM}) and irreversible strain (ε_{ir}) are measured as shown in Fig. 3 and their dependences on the residual strain are presented in Fig. 4. It is seen that an increase in the residual strain increases the shape memory effect value to 2 %. The shape memory effect value does not attain the saturation but the deformation to a larger given strain is not possible due to the destruction of the NiTi layer during bending. The irreversible strain is observed after heating of the sample pre-deformed by 2.8% which shows that plastic strain as well as destruction occur during bending.

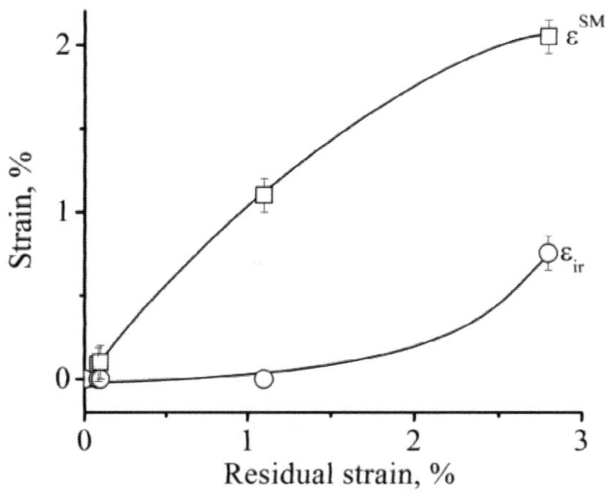

Figure 4. Dependences of the shape memory effect and irreversible strain values on the residual strain.

Thus, NiTi nanofilm deposited to Kapton by flash evaporation demonstrates the one-way shape memory effect however, it does not reveal the two-way shape memory effect. It is known that the two-way shape memory effect is caused by the internal stress fields [17]. Such stress may be induced by the plastic strain however, the irreversible strain is small and found only in the sample with a residual strain of 2.8 %. On the other hand, the internal oriented stress may be accumulated in the

IOP Conf. Series: Materials Science and Engineering 1213 (2022) 012001 doi:10.1088/1757-899X/1213/1/012001

Kapton substrate during the realization of the one-way shape memory effect due to various mechanisms of the strain variation in NiTi film and Kapton, as it is observed in bimetallic shape memory composites [18–20]. However, the results of the study show that internal stress does not appear because the two-way shape memory effect is not observed. This may be caused by the inelastic deformation of Kapton on heating (during the realization of the one-way shape memory effect) that prevents the stress formation but leads to the observation of irreversible strain.

4. Summary

The one-way and two-way shape memory effects were studied in the NiTi films deposited on Kapton substrate and subjected to heat treatment at 350 °C for 2 hours. It was shown that the NiTi/Kapton samples demonstrate the one-way shape memory effect with the maximum recoverable strain of 2%. The two-way shape memory effect was not found in NiTi films deposited on Kapton due to inelastic deformation of Kapton that prevented the formation of internal stress.

5. Acknowledgments

This work was supported by joint DST-RSF project (RSF # 19-49-02014, DST #DST/INT/RUS/RSF/P-36). The X-ray and SEM studies were carried out using the equipment of X-ray Diffraction Centre and Resource Centre for Nanotechnology of Saint Petersburg State University. The authors thanks Doctor I.V. Kazakov (SPbSU) for his assistance in samples vacuumization.

References

[1] Ameduri S, Concilio A, Favaloro N and Pellone L 2016 *Aerospace* **3** 16
[2] Ameduri S and Concilio A 2019 *J. Intell. Mater. Syst. Struct.* **30** 2605–2626
[3] Mohd Jani J, Leary M, Subic A and Gibson M A 2014 *Mater. Des.* **56** 1078–1113
[4] Sibirev A, Belyaev S and Resnina N 2021 *Sensors Actuators A Phys.* **319**
[5] Shin D D, Mohanchandra K P and Carman G P 2005 *Sensors Actuators, A Phys.* **119** 151–156
[6] Makino E, Uenoyama M and Shibata T 1998 *Sensors Actuators, A Phys.* **71** 187–192
[7] Mineta T, Kasai K, Sasaki Y, Makino E, Kawashima T and Shibata T 2009 *Microelectron. Eng.* **86** 1274–1277
[8] Busch J D, Johnson A D, Lee C H and Stevenson D A 1990 *J. Appl. Phys.* **68** 6224–6228
[9] Fu Y and Du H 2002 *Surf. Coatings Technol.* **153** 100–105
[10] Kim S W, Jeon Y M, Park C H, Kim J H, Kim D H and Yeom J T 2013 *J. Alloys Compd.* **580** 5–9
[11] Bechtold C, Chluba C, Lima de Miranda R and Quandt E 2012 *Appl. Phys. Lett.* **101** 091903
[12] Ishida A and Sato M 2008 *Thin Solid Films* **516** 7836–7839
[13] Jayachandran S, Akash K, Prabu S S M, Manikandan M and Muralidharan M 2019 *Compos. Part B* **176** 107182
[14] Gao Z Y, Wang H Z, Zhu Y Y, Meng X L and Cai W 2015 *Thin Solid Films* **584** 369–371
[15] Karna P, Prabu S S M, Karthikeyan S C, Mithun R, Jayachandran S, Resnina N, Belyaev S and Palani I A 2021 *Sensors Actuators A. Phys.* **321** 112411
[16] Hu C Z and Andrade J D 1985 *J. Appl. Polym. Sci.* **30** 4409–4415
[17] Liu Y and McCormick P G 1990 *Acta Met. Mater* **38** 1321–1326
[18] Belyaev S, Rubanik V, Resnina N, Rubanik V and Lomakin I 2014 *Smart Mater. Struct.* **23**
[19] Belyaev S, Rubanik V, Resnina N, Rubanik V, Lomakin I and Demidova E 2017 *Mater. Today Proc.* **4** 4696–4701
[20] Belyaev S, Rubanik V, Resnina N, Rubanik V, Rubanik O, Borisov V and Lomakin I 2010 *Phys. Procedia* **10** 52–57

Influence of annealing on the functional properties of the NiTi alloy produced by wire arc additive manufacturing

N Resnina[1,3], I A Palani[2], S Belyaev[1], R Bikbaev[1], Shalini Singh[2], S S Mani Prabu[2] and S Jayachandran[2]

[1]Saint-Petersburg State University, Saint-Petersburg, Russia
[2]Department of Mechanical Engineering, Indian Institute of Technology Indore, Indore, India

[3]Corresponding author's e-mail: resnat@mail.ru

Abstract. The influence of the annealing temperature on the recoverable strain variation on cooling and heating under a stress of 200 MPa was studied in the NiTi samples produced by wire arc additive manufacturing. The samples including the Ni-rich NiTi layer in the working length were annealed for 10 hours at various temperature from 450 to 600 °C. It is shown that an increase in annealing temperature leads to non-monontonic variation in the recoverable strain. This is caused by an increase in annealing temperature from 450 to 550 °C increases the volume fraction of Ni_4Ti_3 precipitates. As a result, the volume fraction of the NiTi phase undergoing the martensitic transformation and recoverable strain decrease. An increase in annealing temperature from 550 to 600 °C leads to a dissolving the Ni_4Ti_3 precipitates and formation of the Ni_3Ti_2 precipitates that increases the volume fraction of the NiTi phase and the recoverable strain.

1. Introduction

Shape memory alloys are commercially used in different application fields such as aerospace, mechanical engineering, medicine and robotics due to their unusual deformation behaviour [1-2]. These materials demonstrate the strain recovery on heating (the one-way shape memory effect) or unloading (the superelasticity) of the pre-deformed sample, the recoverable strain variation on cooling and heating under a stress or without stress (the two-way shape memory effect) or recovery stress generation on heating under strain-fixed state [3]. Shape memory alloys include different alloy systems but the NiTi-based alloys demonstrate the best combination of functional and mechanical properties. At the same time, the production of the NiTi-based massive elements with complex shape is a challenging procedure due to these materials are characterised by a weak workability. Using the additive manufacturing technology is the promising way to solve this problem.

The bed deposition technology is effectively used for the fabrication of small NiTi parts while the direct deposition technology is convenient for the production of massive elements [4]. One of the direct deposition technology is the wire arc additive manufacturing (WAAM) [5-6] which now is used for the production of NiTi samples including walls and 3D structures [7-11]. Previously, it was found that the structure of the NiTi samples produced by WAAM depended on the substrate materials, arc voltage, arc current and pre-heating temperature of substrate [12-14]. Regardless of these parameters, the common feature of the NiTi samples produced by WAAM is the different chemical composition of the NiTi phase along the sample height (built direction) due to the penetration of the substrate elements to the NiTi melt [10]. This leads to the martensitic transformations occur at various temperatures in different layers of the sample [10]. As a result, the strain recovery is observed in a wide temperature range that is not convenient for application [10]. One of the tools for the control of the chemical composition of the NiTi phase is heat treatment that allows to change the chemical composition of the NiTi phase, type and size of precipitates and affect the martensitic transformation [3]. The influence of the heat treatment on the structure and the martensitic transformation in the 5-layerd NiTi walls has been studied previously and it was found that the annealing for 10 hours at a temperature belonged to a range of 450 – 600 °C allowed to align the chemical concentration of the NiTi phase within the layer but kept the different concentration in various layers. At the same time, the chemical composition of the NiTi phase, precipitates and the parameters of the martensitic transformation affect the functional behaviour of the sample hence it should by depended on annealing temperature but it has not been investigated in the NiTi sample produced by WAAM. Thus, the aim of the present paper is to study the influence of the annealing temperature on the recoverable strain variation in NiTi samples produced by WAAM.

2. Materials and methods

A 5-layered sample was deposited on a titanium substrate by WAAM using a Ti-50.9 at % Ni wire with a diameter of 1.2 mm (details are given in [10]). The bulk sample was cut by electro-discharge machine into cross-sections plates with a thickness of 0.5 mm, which were used to cut the sample for tension which working length of 6 mm included the 3rd and 4th layers (a working width was 1 mm). The samples were annealed at 450 °C, 500 °C, 550 °C and 600 °C for 10 hours and quenched into water. The structure and the martensitic transformations have been studied in each layer previously and the summarizing results are given in Table 1.

Table 1. Structure and parameters of martensite transformations in the 3rd and 4th layers after various heat treatments for 10 hours. Low indexes: s – start temperature; f – finish temperature. MT – martensitic transformation.

Annealing temperature (°C)	Layer	Structure	Forward MT sequence	Forward MT T (°C) R_s	R_f	M_s	M_f	Reverse MT sequence	Reverse MT T (°C) A_s	A_f
450	3	$NiTi+Ti_2Ni$	B2 →B19'	-	-	37	16	B19'→ B2	47	62
	4	$NiTi+Ni_4Ti_3+Ti_2Ni$	B2 → R → B19'	41	29	8	-14		36	46
500	3	$NiTi+Ti_2Ni$	B2 →B19'	-	-	34	13		35	59
	4	$NiTi+Ni_4Ti_3+Ti_2Ni$	B2 → R → B19'	25	18	8	-9		31	40
550	3	$NiTi+Ti_2Ni$	B2 →B19'	-	-	32	-4		24	62
	4	$NiTi+Ni_4Ti_3+Ti_2Ni$	B2 → R → B19'	12	-	-2	-14		21	26
600	3	$NiTi+Ti_2Ni$	B2 →B19'	-	-	-14	-73		11	43
	4	$NiTi+Ni_3Ti_2+Ti_2Ni$	B2 →B19'	-	-	-57	-74		-44	-31

To study the strain variation under tensile mode, special inter-grippers were used to hold sample for tension to standard flat grips of the Lloyd 30k Plus testing machine (see details in [10]). The strain was measured by video extensometer as the variation in the gap between two white marks. To find the influence of the annealing temperature on the functional properties of the NiTi sample produced by WAAM, the recoverable strain variation on cooling and heating under a stress of 200 MPa was studied. The sample was heated to temperatures higher then finish temperature (A_f) of the reverse martensitic transformation in the 3rd layer at which both 3rd and 4th layers were in the austenite state, loaded to a stress 200 MPa and subjected to cooling and heating through a temperature range of the martensitic transformation. The loading was carried out under the displacement control at a rate of 0.02 mm/min. The samples annealed at 450 – 550 °C were cooled and heated in a temperature range from 200 °C to -15 °C. The sample annealed at 600 °C was cooled and heated in a temperature range from 200 °C to -80 °C. The temperature variation during cooling and heating was 5 °C/min.

3. Results and Discussion

Fig. 1a shows the ε(T) curves found on cooling and heating under 200 MPa in the samples annealed at different temperatures. Two-stage strain variation is found on cooling and heating of all samples under a stress however, the larger the annealing temperature the larger the temperature range of the first stage on cooling and last stage on heating. The observation of various stages on the ε(T) curve is cause by the difference in transformation temperatures in the 3rd and 4th layers and various influence of annealing temperature on the martensitic transformation (see Table 1). On increase in the annealing temperature, the difference in transformation temperatures for the 3rd and 4th layers increases (see Table 1) that leads separation of the temperature ranges of strain variation induced by the transformation in these layers. An increase in the annealing temperature decreases the temperatures of strain variation that is in good agreement with the influence of the annealing on the transformation temperatures (see Table 1).

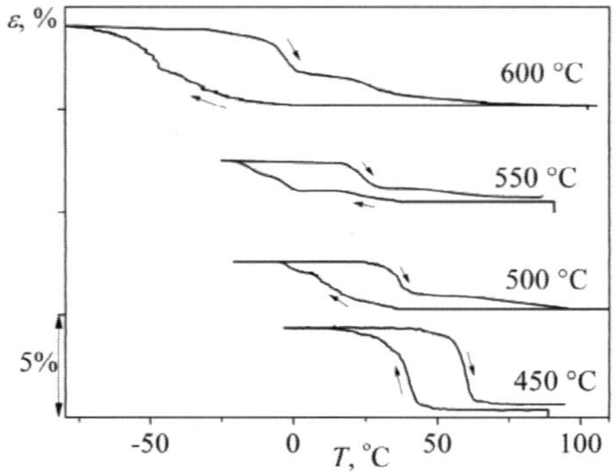

Figure 1. Strain variation found on cooling and heating under a 200 MPa of the NiTi sample annealed at different temperatures for 10 hours.

The recoverable strain (ε^{SM}) has been measured as shown in Fig. 2a and their dependence on the annealing temperature is given in Fig. 2b. It is seen that an increase in the annealing temperature from 450 to 550 °C decreases the ε^{SM} value from 3.8 to 1.8 %. Further increase in annealing temperature to 600 °C increases the shape memory effect value up to 3.8 %. This is caused by an increase in annealing temperature from 450 to 550 °C leads to an increase in the size of the Ni_4Ti_3 precipitates that decreases the volume of the NiTi phase. As a results, the value of the recoverable strain decreases. During annealing at 600 °C, the Ti_3Ni_4 particles dissolve and Ni_3Ti_2 particles form but their volume fraction is less than the volume fraction of Ni_4Ti_3 precipitates in the sample annealed at 450 – 550 °C [3]. In this case, the volume fraction of the NiTi phase increases and the recoverable strain rises. The irrecoverable strain is found in the sample annealed at 450 or 550 °C and its value was 0.2 ±0.1 %.

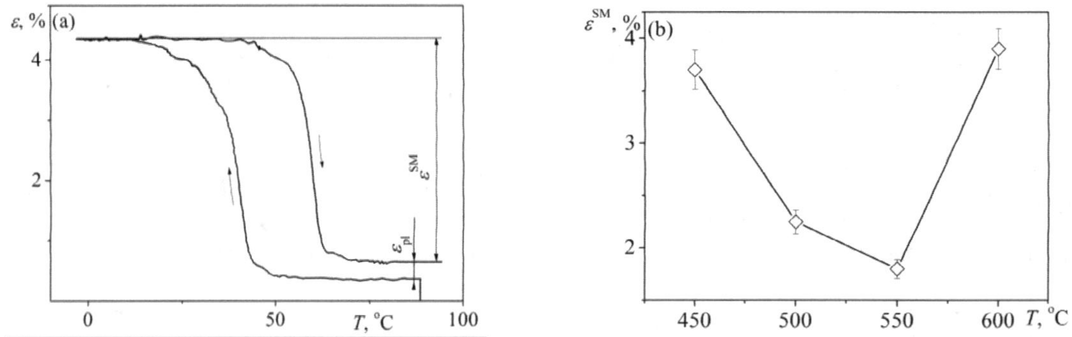

Figure 2. The ε(T) curves found on cooling and heating under a stress of 200 MPa of the sample annealed at 450 °C for 10 hours (a) and the influence of the annealing temperatures on the recoverable strain (b).

4. Conclusions

The influence of annealing temperature on the recoverable strain variation on cooling and heating under 200 MPa was studied in the NiTi sample produced by WAAM. It was found that an increase in annealing temperature from 450 to 600 °C led to non-monontonic variation in the recoverable strain. This was caused by an increase in annealing temperature from 450 to 550 °C increased the volume fraction of Ni_4Ti_3 precipitates that decreased the volume fraction of the NiTi phase undergoing the martensitic transformation. As a result, the recoverable strain decreased on a rise in the annealing temperature. An increase in annealing temperature from 550 to 600 °C resulted in a dissolving the Ni_4Ti_3 precipitates and formation of the Ni_3Ti_2 precipitates that increased the volume fraction of the NiTi phase undergoing the martensitic transformation and the recoverable strain increased.

Acknowledgements
This work was supported by joint DST-RSF project (RSF # 19-49-02014, DST #DST/INT/RUS/RSF/P-36).

References

[1] Razov AI 2004 *Phys. Met. Metallogr.* **97** 97-126
[2] Jani JM, Leary M, Subic A, Gibson MA 2014 *Mater. Des.* **56** 1078-1113
[3] Otsuka K, Ren X 2005 *Prog. Mater. Sci.* **50** 511-678
[4] DebRoy T, Wei HL, Zuback JS, Mukherjee T, Elmer JW, Milewski JO, Beese AM, Wilson-Heid A, De A, Zhang W 2018 *Prog. Mater. Sci.* **92** 112–224
[5] Liu J, Xu Y, Ge Y, Hou Z, Chen S 2020 *Int. J. Adv. Manuf. Tech.* **111** 149–198
[6] Wu B, Pan Z, Ding D, Cuiuri D, Li H, Xu J, Norrish J 2018 *J. Manuf. Proces.* **35** 127–139
[7] Zeng Z, Cong BQ, Oliveira JP, Ke WC, Schell N, Peng B, Qi ZW, Ge FG, Zhang W, Ao SS 2020 *Addit. Manuf.* **32** 101051
[8] Wang J, Pan Z, Wang Y, Wang L, Su L, Cuiuri D, Zhao Y, Li H 2020 *Addit. Manuf.* **34** 101240
[9] Resnina NN, Palani IA, Liulchak PS, Belyaev SP, Mani Prabu SS, Jayachandran S, Kalganov V 2020 *Lett. Mater.* **10** 496–500
[10] Resnina N, Palani IA, Belyaev S, Mani Prabu SS, Liulchak P, Karaseva U, Manikandan M, Jayachandran S, Bryukhanova V, Anshu Sahu, Bikbaev RM 2021 *J. Alloys Compd.* **851** 156851.
[11] Singh S, Resnina N, Belyaev S, Jinoop AN, Shukla A, Palani IA, Paul CP, Bindra KS 2021 *J. Manuf. Proces.* **66** 70–80
[12] Wang J, Pan Z, Wang Y, Carpenter K, Li H 2020 *in B. Mrugalska et al. (Eds.): AHFE* **1216** 78–85
[13] Wang J, Pan Z, Carpenter K, Han J, Wang Z, Li H 2021 *Mat. Sci. Eng. A* **800** 140307
[14] Ponikarova I, Palani IA, Liulchak P, Resnina N, Singh S, Belyaev S, Mani Prabu SS, Jayachandran S, Kalganov V, Anshu Sahu, Bikbaev R, Karaseva U 2021 *J. Manuf. Proces.* **70** 132-139.
[15] Resnina N, Palani IA, Belyaev S, Singh S, Mani Prabu SS, Bikbaev R, Jayachandran S, Kalganov V 2021, *Mater. Lett.* **298** 130004
[16] Belyaev S, Resnina N, Rakhimov T, Andreev V 2020 *Sensors and Actuators A* **305** 111911

IOP Conf. Series: Materials Science and Engineering 1213 (2022) 012003 doi:10.1088/1757-899X/1213/1/012003

Slippage during high-pressure torsion of commercially pure titanium and application of accumulative HPT to it

D V Gunderov*[1,2], A A Churakova[1,2], A V Sharafutdinov[2], V D Sitdikov[2], V V Astanin[2]

[1] Institute of Molecule and Crystal Physics - Subdivision of the Ufa Federal Research Center of the Russian Academy of Sciences, 151 pr. Oktyabrya, Ufa, 450075, Russia
[2] Ufa State Aviation Technical University, 12 K. Marx str., Ufa, 450008, Russia

*Corresponding author e-mail address: dimagun@mail.ru

Abstract. A new efficient method was used to find that in the case of high-pressure torsion of commercially pure titanium, accumulation of shear strain in Ti does not occur due to slippage of anvils. Despite this, micro-hardness increases as the number of turns n increases, and Ti structure is refined more intensively. High-pressure torsion is accompanied by a high-pressure ω-phase formation. However, the content of ω-phase changes non-monotonously with an increase in the number of turns. First, while number of turns is less than n=5, the ω-phase content reaches 50%. Upon further deformation, the ω-phase content decreases to 15% for n=20. A new accumulative high-pressure torsion method is applied to commercially pure titanium for the first time. Accumulative high-pressure torsion leads to the strongest transformation of the structure and an increase in hardness, since stronger real deformation occurs due to composition of compression and torsion strain cycles.

1. Introduction

Titanium (Ti) and its alloys stand out among the available materials for various applications including the production of medical implants [1]. An additional increase in the strength characteristics of titanium and its alloys may be achieved through the formation of a nanostructured state by severe plastic deformation [2]. High-pressure torsion (HPT) in an intensive plastic strain method that allows refining the structure of metallic materials to the maximum possible extent by forming a nanostructured state in materials [2]. The researchers use anvils of various geometry, various pressures and temperatures, number of revolutions, anvil rotation speed, and the initial state of the material subjected to HPT, so the results of various authors for HPT of the same material can significantly differ [1–7].

A large number of articles are devoted to the study of HPT effects on pure titatium [3–9]. It was shown that HPT of pure titanium led to a structure refined to the grain size of 80 nm. The achieved minimal grain size varies considerably among different authors, which is related to various HPT conditions and titanium grades and impurities [3–10]. The microhardness HV increases with an increase in the equivalent strain at an early-stage HPT. However, upon reaching a certain number of revolutions of the anvils (usually n=5), HV value (and structural change) reaches limit and remain the same after some critical strain by HPT [10]. So, the maximum limit of hardness depends on applied pressure P. Pure titanium in the initial state at room temperature has α hcp crystal structure, which transforms upon deformation under high pressure to a ω phase with a simple hexagonal structure [11]. The studies [6,8] report that the formation of the ω phase occurred in titanium during HPT even at a pressure of 3–4 GPa. HPT deformation facilitates the α – ω phase transformation [6,8], and ductility of titanium decreases with the α – ω phase transformation [8].

Strain γ achieved during HPT can be calculated using the formula [1]:

$$\gamma = \frac{2\pi Rn}{h}, \quad (1)$$

where R – radius of the measurement point, n – number of revolutions of anvils, h – thickness of the specimen.

However, it is known for HPT of hard and hardened materials, anvils may slip along the specimen surface and real strain γ_{real} does not correspond to calculated using (1). Anvil slippage at HPT is shown in many papers, in particular, [12–15]. Bridgeman shows that during torsional deformation slippage between the specimen and the anvil surface is possible due to insufficient friction force [13]. Slippage is low at HPT of aluminum and copper, however, it becomes much greater during HPT of iron [12]. In [12], the authors evaluated the slippage by applying marks on the upper and lower surfaces of the disk subjected to HPT. For HPT with θ>90°, these marks were erased, which prevented from evaluating strain based on the relative shear of marks [12]. The paper [15] proposed a new method of joint deformation of the halves to evaluate the slippage degree at HPT with which it was shown that shear strain γ in hard bulk metallic glass (BMG) based on Zr (YTS>1500 MPa) at HPT with n=0.25–5 was much lower than expected, and while there is no shear strain at HPT, the structure of BMG significantly changed. In [16], slippage at HPT Zr-1%Nb alloy was studied, and it was shown that the specimen of Zr-1%Nb at the initial HPT stages (n=1, n=2) received a significant degree of torsional deformation, However, after HPT with more than 5 revolutions, torsional deformation was not implemented due to slippage. After HPT n=10 the structure of the same alloy is slightly finer, but the defect density is much higher compared to the same alloys after HPT n=5. Hence, it can be assumed that despite slippage during HPT, the sample receives deformation even when revolution number is increased to n>5. Establishing the mechanisms of this phenomenon require additional research. Currently, the method of [15] to evaluate slippage at HPT is not widely used for metals and alloys.

In [17], the new method was used for the first time – so-called accumulative high-pressure torsion (ACC HPT). For ACC HPT, the strain is accumulated by combining the cycles of HPT, specimen cutting, placing quarters on anvils, upsetting of quarters, and then HPT cycle again (this method is briefly described in "Material and methods"). During cycles in upsetting and at the initial stage of torsion, the specimen receives significant strain. Such cycles can be repeated several times so high strains can be achieved even for hard and work-hardened materials. As said above, for usual HPT conditions, when n rises in such materials, anvils start to slip. Previously ACC HPT was applied to steel 316, β-Ti, Zr-1%Nb [16,18]. It was shown that ACC HPT led to a stronger refinement of the structure of these materials than conventional HPT. In this study, ACC HPT is applied to commercially pure titanium (CPT).

2. Materials and methods

This study used commercially pure titanium of Russian grade VT1-0, roughly equivalent to Titanium Grade 1. The chemical composition of VT1-0 according to GOST 19807-91 is given in Table 1.

Table 1. Chemical composition of VT1-0 titanium, in wt. %.

Ti	Al	Si	Fe	C	O	N	H	Residuals
balance	<0.70	<0.10	<0.25	<0.07	<0.20	<0.04	<0.010	<0.3

To carry out deformation by high pressure torsion, disks with a diameter of 10 mm and a thickness of up to 1 mm were cut from the initial material. Then the samples were subjected to HPT at room temperature on anvils 10 mm in diameter and a pressure of 6 GPa for n=5, 10, 20 revolutions.

The method of finding the degree of strain at HPT was earlier described in [15], and is as follows: the initial metal disc was cut into two segments. The edges of segments were polished and varnished to avoid metal adhesion. The combined Ti segments were then subjected to HPT.

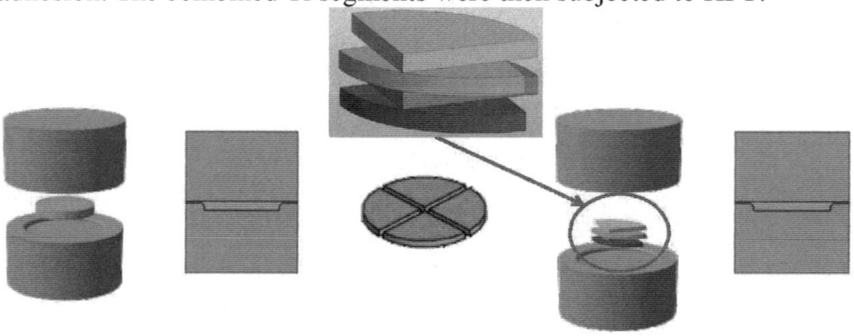

Figure 1. Diagram of accumulative high-pressure torsion.

Figure 1 shows a diagram of the accumulative HPT process. The titanium specimen was subjected to HPT under the scheme n=1+1+1+1+1+5+1+1+1+1+1+5, e.g., anvils were rotated once, then the specimen was broken into 4 pieces placed on each other on the anvils, which were rotated once again (n=1). Five such cycles were completed (breaking-placing+1 revolution) with the 6th cycle being breaking-placing+5 revolutions. Then 5 such cycles were repeated (breaking-placing+1 revolution) with the 12th cycle being breaking-placing+5 revolutions. A total of 20 revolutions were made, $n_{\Sigma}=20$. As a result, the material received high strain, since the sample is deformed during upsetting and at the initial stage of torsion.

Micro-hardness was measured using the Vickers method under load 1 N (100 g) for 10 sec. The X-ray diffraction (XRD) analysis was carried out on a Rigaku Ultima IV diffractometer. The samples were examined with CuKα-radiation (40 kV, 30mA) and the phase composition of the alloy was determined using the Rietveld method.

3. Results and discussion

To determine the real achieved strain by torsion, two Ti segments were subjected to joint HPT by a quarter of a turn according to the scheme (Figure 2, a) [16]. The specimen became of complex shape, but no significant shear of the upper and lower surfaces of the specimen was found. Let us note that in the same experiment with Cu, the appearance of Cu segments after combined HPT at θ=90° displaced relative to each other by θ=90° [16]. Hence, the real shear strain Ti after HPT n=1/4 was much lower than $\gamma_1=8$, which is predicted by formula (1). This is explained by the fact that significant slippage occurs at the initial stage of HPT of Ti. The revealed significant slippage at HPT of Ti shows the need for using accumulative HPT to achieve an increased degree of deformation on Ti.

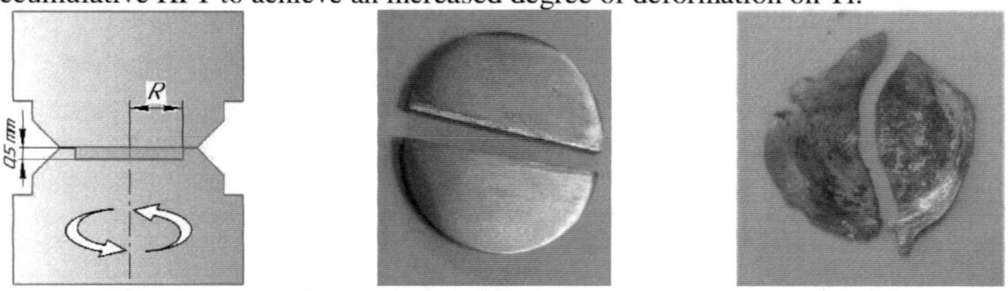

a) b) c)

Figure 2. Scheme of combined HPT of two segments to find the degree of shear strain [16] (a), two segments of the Ti disc, prepared for joint HPT (b), segments after joint HPT n= ¼ (c).

After HPT, micro-hardness of Ti more than doubles (Table 2), which proves a strong refinement of the structure. HV values after HPT n=5, n=10 are lower in the center of specimens as compared with the half radius (1/2R). This is a known result related to a lower degree of strain in the center of the specimen. As the number of revolutions reaches n=20, HV values in the center of the specimens increase even more, but in the area of 1/2R, HV reaches maximum at n=5. For HPT n=20, HV values in the specimen center are higher than in the area of 1/2R. Possibly, this is related to the heavily reduced thickness of the specimen in the area of 1/2R after HPT n=20. The highest increase of HV takes place after ACC HPT, which proves the highest introduced strain to the material. Hence, the accumulating strain leads to a higher real strain and additional structure refinement.

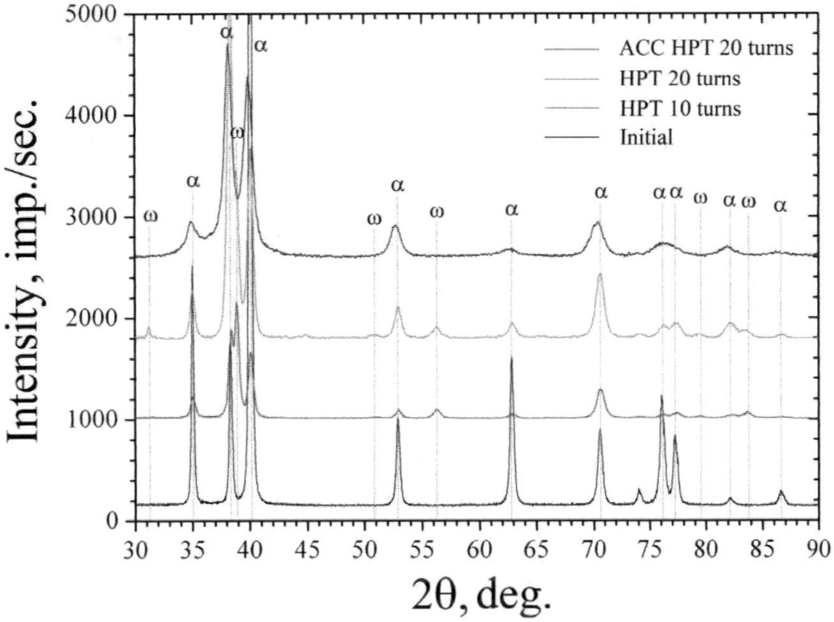

Figure 3. X-ray diffraction profiles of the Ti samples: a) initial; b) subjected to HPT n=10; c) subjected to HPT n=20; d) subjected to ACC HPT n_Σ=20.

Table 2. Results of microhardness measurements, HV, grid distortion, in %, coherent scattering region size, in nm, and ω-phase content, in %, according to XRD analysis of Ti before and after HPT.

Condition	Position	HV0.1	Grid distortion, %	CSR, nm	share of ω-phase, %
initial	-	210	0.17	72	0
HPT n=5	center 1/2R	455 475	0.25	23	46
HPT n=10	center 1/2R	440 450	0.34	16	20
HPT n=20	center 1/2R	470 390	0.48	27	15
ACC HPT n_Σ=20	center 1/2R	570 515	0.61	13	2

After HPT in all conditions, the ω-phase appears due to α→ω transition as earlier shown in [6,12]. After HPT n=5, the content of the ω-phase reaches 50%. However, as the number of anvil revolutions further rises at HPT, the content of the ω-phase falls to 15% n=20. The decreased content of the ω-phase above some critical value as the number of anvil revolutions increases was also noted by other authors [6,12]. The reconstructive analysis also showed that HPT resulted in growing micro-distortions and the size of coherent scattering region (CSR) decreased. The highest grid distortions and the lower size of coherent scattering region are achieved as a result of ACC HPT. This also proves that ACC HPT leads to a stronger refinement of the structure due to composition during shrinkage and torsion stain.

4. Conclusions

To find the actually achieved degree of strain by torsion, two Ti segments were subject to combined HPT by n=1/4. The experiment showed that the real shear strain Ti after HPT n=1/4 is much lower than predicted. This is explained by the fact that significant slippage occurs at the initial stage of HPT of Ti. The result supports the relevance of using accumulative HPT. HPT results in growing Ti micro-hardness. The highest increase in micro-hardness takes place after ACC HPT, which indicates the highest strain applied to the material. After HPT in all conditions, ω-phase appears due to α→ω transition. After HPT of n=5, the ω-phase content reaches 50% but, as n further rises, the ω-phase content falls down. XRD showed that the highest grid distortions and the lower size of coherent scattering regions were achieved as a result of ACC HPT. This also proves that ACC HPT leads to a stronger refinement of the structure since the highest real strain is achieved due to composition during shrinkage and torsion stain.

Acknowledgements

The work is supported by the Russian Foundation for Basic Research, Project No. 19-58-80018. The authors are grateful to the personnel of the research and technology Joint Research Center, Nanotech, Ufa State Aviation Technical University (http://nanotech.ugatu.ac.ru/) for their assistance with instrumental analysis.

References

[1] Valiev R Z, Zhilyaev A P and Langdon T G 2013 *Bulk Nanostructured Materials* (Hoboken, NJ: John Wiley & Sons, Inc)
[2] Zhilyaev A P and Langdon T G 2008 *Prog. Mater. Sci.* **53** 893–979
[3] Popov A A, Pyshmintsev I Y, Demakov S L, Illarionov A G, Lowe T C, Sergeyeva A V and Valiev R Z 1997 *Scr. Mater.* **37** 1089–94
[4] Valiev R Z, Sergueeva A V. and Mukherjee A K 2003 *Scr. Mater.* **49** 669–74
[5] Stolyarov V ., Zhu Y ., Lowe T ., Islamgaliev R . and Valiev R . 1999 *Nanostructured Mater.* **11** 947–54
[6] Kilmametov A R, Khristoforov A V., Wilde G and Valiev R Z 2007 *Zeitschrift für Krist.* **2007** 339–44
[7] Islamgaliev R K, Kazyhanov V U, Shestakova L O, Sharafutdinov A V and Valiev R Z 2008 *Mater. Sci. Eng. A* **493** 190–4
[8] Edalati K, Matsubara E and Horita Z 2009 *Metall. Mater. Trans. A* **40** 2079–86
[9] Wang C T, Escudeiro A, Polcar T, Cavaleiro A, Wood R J K, Gao N and Langdon T G 2013 *Wear* **306** 304–10
[10] Shirooyeh M, Xu J and Langdon T G 2014 *Mater. Sci. Eng. A* **614** 223–31
[11] Sikka S K, Vohra Y K and Chidambaram R 1982 *Prog. Mater. Sci.* **27** 245–310
[12] Edalati K, Horita Z and Langdon T G 2009 *Scr. Mater.* **60** 9–12
[13] Bridgman P W 1937 *Proc. Am. Acad. Arts Sci.* **71** 387
[14] Kuhlmann-Wilsdorf D, Cai B C and Nelson R B 1991 *J. Mater. Res.* **6** 2547–64
[15] Gunderov D V, Churakova A A, Boltynjuk E V, Ubyivovk E V, Astanin V V, Asfandiyarov R N, Valiev R Z, Xioang W and Wang J T 2019 *J. Alloys Compd.* **800** 58–63
[16] Gunderov D, Stotskiy A, Lebedev Y and Mukaeva V 2021 *Metals (Basel).* **11** 573
[17] Gunderov D V, Churakova A A, Astanin V V, Asfandiyarov R N, Hahn H and Valiev R Z 2020 *Mater. Lett.* **261** 127000

IOP Conf. Series: Materials Science and Engineering 1213 (2022) 012003 doi:10.1088/1757-899X/1213/1/012003

[18] Gunderov D V, Khasanova D A, Astanin V V, Churakova A A, Ramazanov I A, Prokoshkin S D and Sheremetyev V A 2020 *IOP Conf. Ser. Mater. Sci. Eng.* **1008** 012013

IOP Conf. Series: Materials Science and Engineering 1213 (2022) 012004 doi:10.1088/1757-899X/1213/1/012004

Influence of the initial structural-phase state on inelastic and plastic strains behavior in coarse-grained samples of the $Ti_{49.3}Ni_{50.7}$ (at.%) alloy

D Yu Zhapova[1], A I Lotkov[1], V N Grishkov[1], A A Gusarenko[1], I S Rodionov[1]

[1] Institute of Strength Physics and Materials Science of SB RAS, 2/4, Akademicheskii av., Tomsk, 634055, Russia

*Corresponding author: dorzh@ispms.tsc.ru

Abstract. The paper presents the experimental results of studies of the temperature dependence of inelastic and plastic strains during torsion of coarse-grained samples of the $Ti_{49.3}Ni_{50.7}$ (at.%) alloy. Investigations of the deformation behavior of the test alloy samples in the martensitic, two-phase and high-temperature states have been carried out. It is shown that the value of the summary inelastic strain reaches a maximum value of ~ 18% under deformation of the samples in the martensitic and two-phase state, as well as in the temperature range of pre-transition phenomena.

1. Introduction

Titanium nickelide (TiNi) based alloys are applied in various technology and medicine fields as a functional material with superelasticity and shape memory effects (SME) [1-3]. These effects are due to thermoelastic martensitic transformations (MTs) from a high-temperature B2 phase to a monoclinic martensitic phase B19' [1, 4-7]. The manifestations of reversible inelastic deformation are determined by the position of the temperature of the isothermal loading-unloading cycles, T_{def}, relative to the temperature ranges of the MT realization in the initial unloaded samples [5, 7-9]. Both during the realization of the superelasticity effect and during the realization of the shape memory effect (SME), the samples of TiNi - based alloys undergo deformation effects. During the realization of superelasticity effect, isothermal loading is carried out to a certain total deformation, followed by unloading at the same temperature. The traditional scheme of the SME realization includes: samples cooling to the deformation temperature, T_{def} (below the temperature of the end of the MT into the B19' phase) without applying a load, isothermal loading up to a total strain and unloading followed by heating of the unloaded samples until the completion of the reverse MT and, accordingly, the completion of the shape recovery of samples [5, 7]. Thus, the temperature of isothermal loading-unloading cycles (T_{def}) and the value of the total strain are important parameters that can affect the manifestation of inelastic and plastic strains in alloys with thermoelastic MTs. Therefore, the study of the influence of T_{def} on the manifestation of the superelasticity, SME and summary inelastic strain of samples is a current problem in the design of technical and medical products from TiNi based alloys.

The aim of this paper is to study the regularities of the effect of the isothermal loading temperature on the manifestations of inelastic strain and the development of plastic strain in the $Ti_{49.3}Ni_{50.7}$ (at.%) binary alloy with coarse – grained structure for medical purposes.

2. Materials and Methods

The $Ti_{49.3}Ni_{50.7}$(at.%) alloy (MATEK-SMA Ltd, Russia) was selected as a test material for study. The samples were made from rods (ø 30 mm) obtained by hot rotary forging (additional heat treatment was not carried out). The structural-phase state of the samples was investigated by the X-ray diffraction method at room temperature (Co-Kα radiation) using a DRON-7 setup and by optical metallography using an AXIOVERT-200 MAT optical microscope (Carl Zeiss). Studies have shown that at room temperature the samples were in the state of the high-temperature B2 phase. The X-ray diffraction patterns showed that the fraction of $Ti_4Ni_2(O,N,C,H)_x$ phase in samples was less than 5 vol%. The average grain size was 53 ± 11 μm.

The determination of temperatures and the sequence of martensitic transformations (MT) were carried out by the method of temperature resistometry. Samples underwent the B2↔B19' MTs under cooling and heating. The temperatures of the onset and end of the MT into the B19' phase were M_S = 252K and M_F = 223K. The temperatures of the onset and the end of the reverse martensitic transformation into the B2 phase were A_S = 258K and A_F = 273K. The inelastic and plastic strains of the samples were investigated by torsion method on a unit similar to the inverted torsional pendulum with operating temperature range from 120K to 573K. The "τ - γ" dependences were obtained in isothermal cycles "loading-unloading" under torsion of samples and with subsequent heating of unloaded samples to 550K (~ 200 degrees above A_F). In each subsequent "loading-unloading" cycle of the same samples, γ_t was increased up to the destruction of the samples. Isothermal cycles "loading - unloading" were carried out at the following deformation temperatures: 164K (martensitic state), 242K (two-phase state B2 + B19'), 260K and 281K (pre-transition state) and 299K, 315K and 339K (high-temperature B2 state). All components of a total strain γ_t, which sequentially increased from cycle to cycle, were determined when a torsional strain was applied to the samples. The total strain γ_t is the sum of the γ_{SIS} and the plastic strain γ_{rp}, Fig. 1. The inelastic strain γ_{SIS} is equal to the sum of the value of superelasticity effect manifested in the process of isothermal unloading (including Hooke's strain) and the value of the shape memory effect realized when the unloaded samples are heated until the completion of shape recovery, where γ_r is the residual strain after isothermal unloading of the samples. Plastic strain γ_{rp} of samples is equal to residual strain under heating the unloaded samples to 550K, Fig. 1b.

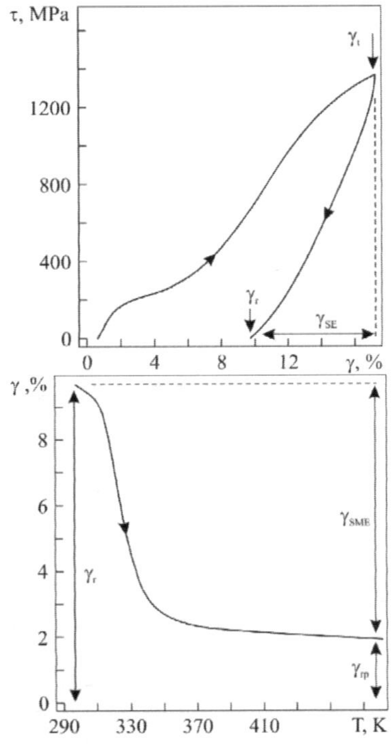

Figure 1. Experimental dependences "τ - γ" under loading and unloading of samples at T_{def} = 299K (a) and under subsequent heating of unloaded samples (b).

3. Results and Discussion

The regularities of the change in the stress-strain curves at different deformation temperatures of the samples are shown in Fig. 2. At the deformation temperature of 164 K, the study samples are in the state of the martensitic phase B19'. The presence of a pseudo-yield plateau is due to the reorientation of the domains of the martensitic phase B19' under the action of external stresses. At T_{def} = 242K and T_{def} = 260K, the reorientation stress of the B19' phase domains decreases to ~ 50 MPa, since the deformation

IOP Conf. Series: Materials Science and Engineering 1213 (2022) 012004 doi:10.1088/1757-899X/1213/1/012004

temperatures are near the temperature of the onset of MT into the B19' phase (M_S = 252K). At T_{def} = 281K (8 degrees above the A_F = 273K), the "τ - γ" dependence is represented a classical dependence typical for TiNi-based alloys with the manifestation of the superelasticity effect and includes: a quasi-elastic stage, a pseudo-yield plateau, a stage of strain hardening with the subsequent transition to the stage of parabolic flow [9]. At T_{def}= 299K, the hardening of samples is observed, and the pseudo-yield plateau becomes less pronounced, Fig. 2 (curve 5). When the samples are deformed at T_{def} = 309K and T_{def} = 339K (curves 6 and 7 in Fig. 2), the pseudo-yield plateau is absent and the quasi-elastic stage immediately passes into the stage of strain hardening.

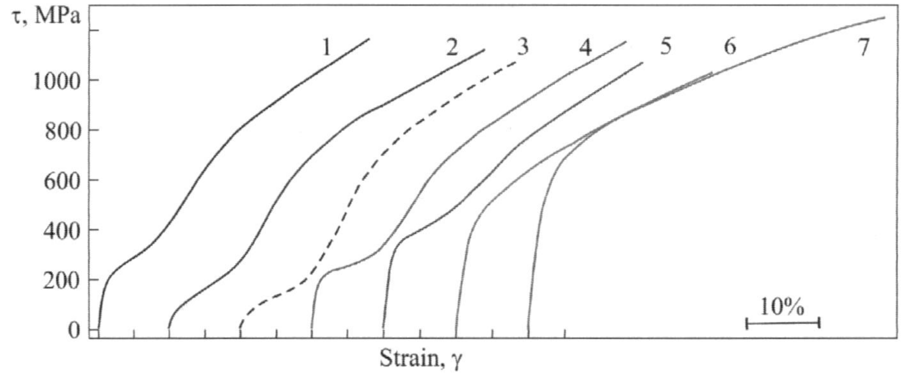

Figure 2. "Stress-strain" dependences under torsion of samples at 164K (1), 242K (2), 260K (3), 299K (5), 309K (6) and 339K (7).

Fig.3 shows the "τ-γ" dependences obtained in the «loading-unloading» cycles with sequentially increasing strains γ_t at T_{def}= 242K and the return of inelastic strain under heating of unloaded samples. It can be seen that the quasi-elastic stage almost immediately passes into the pseudo-yield plateau after the onset of deformation, Fig. 3a. Note that in each deformation cycle, the pseudo-yield plateau can be observed. The return of strain under unloading of samples as effect of superelasticity is small. The main part of the inelastic strain of the samples returns as a shape memory effect when unloaded samples are heated through the temperature range of reverse MTs up to 440 K, Fig. 3b. Similar dependences for T_{def}= 164K are given in [8], and for T_{def}= 281K and T_{def}= 299K in [9].

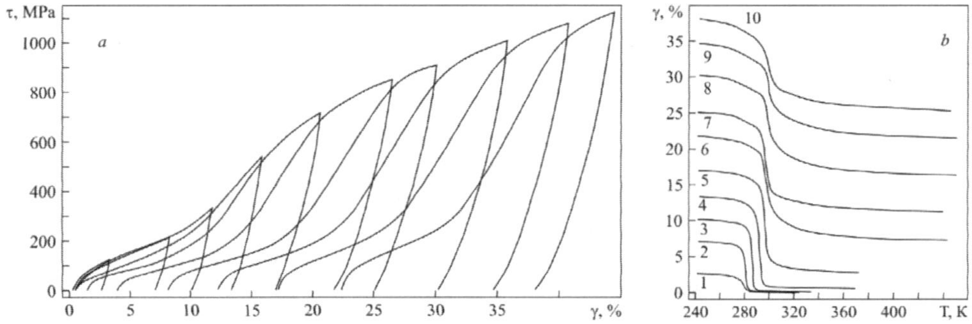

Figure 3. Accumulation and recovery of inelastic strain in isothermal (T_{def} = 242K) loading-unloading cycles (a) and recovery of inelastic strain during intermediate heating of unloaded samples (b).

The regularities of the accumulation and return of inelastic and plastic strains obtained in the "loading-unloading" cycles and during heating of unloaded samples are shown in Fig. 4. It can be seen that at deformation temperatures of 164K and 242K, the return of strain under unloading of the samples is small, which is due to the deformation of the samples in the martensitic state, Figs. 4a and 4b. However, the magnitude of the superelasticity effect reaches 5%. This is possible due to the manifestation of the superelasticity effect caused by reversible twinning of the martensitic phase B19' as discussed earlier in [8]. At T_{def} = 281K, the effect of superelasticity dominates in the first cycle, but in the second cycle of deformation the shape memory effect begins to prevail, Fig. 4c. This behavior is due to the deformation of the samples in the pre-transition temperature range, when the modules of the elastic constants C' and C_{44} decrease in the samples [9]. Under further deformation, defects of the crystal structure accumulate in the samples, which leads to the stabilization of the martensitic phase B19'. For the reverse martensitic transformation of the B19' phase domains, stabilized by crystal structure defects, it is necessary to heat the unloaded samples to temperatures higher than A_F by 10-20 degrees. Therefore, we can observe the shape recovery of the samples under heating. The effect of superelasticity dominates at $T_{def} = 299K$ at any total strains γ_t, Fig. 4d. However, with an increase in the total strain γ_t above 20%, the SME, the value of which can reach 4% is observed. The manifestation of the SME at temperatures above A_F is also due to the stabilization of the martensitic phase B19' by crystal structure defects. Note that, regardless of the deformation temperature, in the first cycles, there is no accumulation of plastic strain (at $T_{def} = 164K$, 242K, and 281K) and its insignificant accumulation is observed when the deformation temperature rises to 299K. Then, a sharp increase in plastic strain is observed, Fig. 4. The value of the summary inelastic strain changes non-monotonically with an increase in the total strain and reaches a maximum value at $\gamma_t \approx 30\%$.

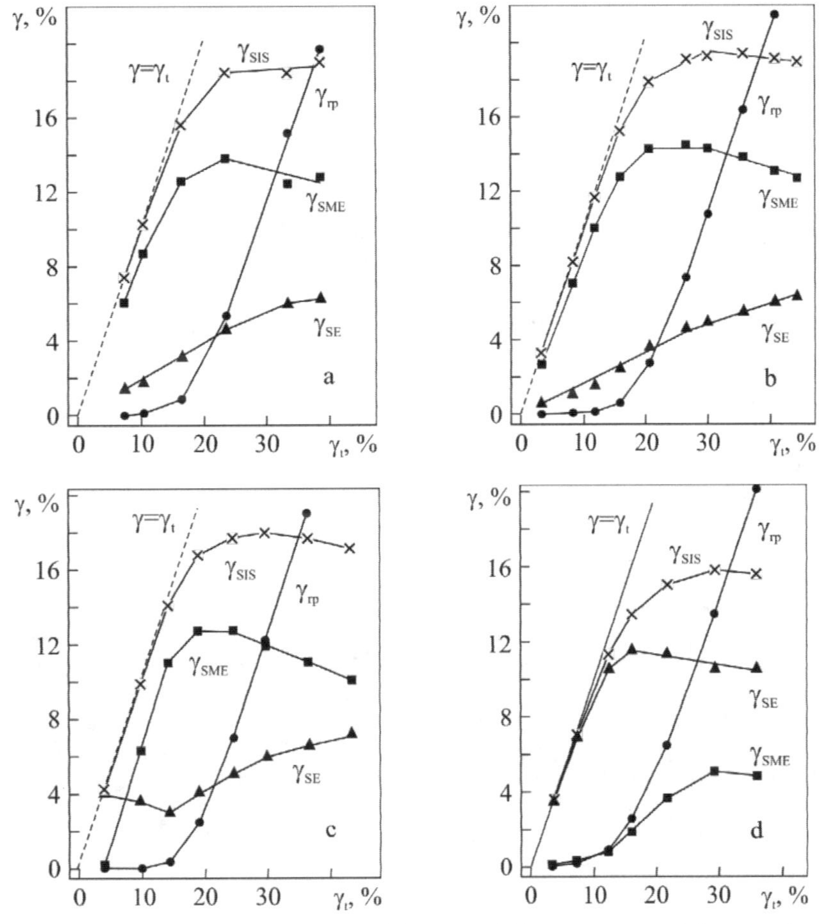

Figure 4. Dependences of γ_t (dotted line), γ_{SE} (▲), γ_{SME} (■), γ_{SIS} (×) и $\gamma_{гр}$ (•) on γ_t after isothermal loading at 164K (a), 242K (b), 281K (c) and 299K (d).

The analysis of dependences of γ_{SE}, γ_{SME}, γ_{SIS} on γ_t obtained in the temperature range from 164K to 229K showed that ~ 98% of the maximum resource of reversible inelastic strain is realized after loading at $\gamma_t = 30\%$. Therefore, the values of the summary inelastic strain at $\gamma_t \approx 30\%$ and the value of $\gamma_{гр}$ corresponding to this total strain at different T_{def} were obtained, Fig. 5. Figure 5 shows that under deformation at $\gamma_t = 30\%$ in the temperature range from 164K to 260K, the samples accumulate virtually the same value of the plastic strain, equal to ~ 12%. Accordingly, the maximum value of γ_{SIS} (~ 18%) is observed in these plastically deformed samples. It can be seen from Fig. 5 that, in process of loading at T_{def}> 299 K, an intense increase in plastic strain γ_{rp} is observed. This leads to degradation of the summary resource of reversible inelastic strain γ_{SIS}.

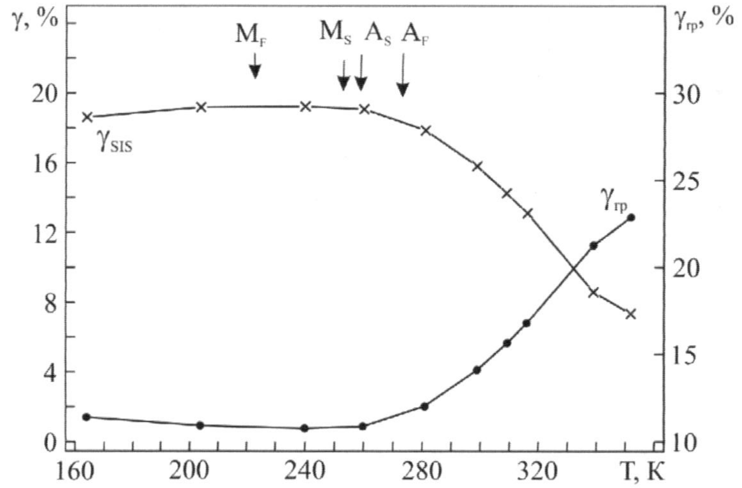

Figure 5. Temperature dependence of the maximum value of the total inelastic strain γ_{SIS} and the corresponding values of the plastic strain γ_{rp}. The value of the total strain γ_t = 30%.

In conclusion, it should be emphasized that the ratio of γ_{SIS} and the plastic strain generated during loading depends on the ratio of the martensitic shear stress (or the stress of reorientation and detwinning of the domains of the martensitic phase B19' at T_{def} <M_F) and the yield stress. According to [10] in the temperature range from 164K to 281K, the difference between the yield stress and the martensitic shear stress is maximum and exceeds 290 MPa, and at T_{def} above 299K this value decreases to ~ 50 MPa and less. Thus, the temperature dependences of the summary inelastic strain and plastic strain are due to the relationship between the yield stress and martensitic shear stress at different T_{def}.

4. Conclusion

It is shown that, at all deformation temperatures the dependences of γ_{SIS} on γ_t have a qualitatively similar form: they increase with an increase in the total strain of torsion, reaching maximum values, and decrease with a further increase in γ_t. The maximum of γ_{SIS} is achieved in the range of total strains γ_t from 24% to 36% (the value of plastic deformation in this case is equal to 11-23%). In the range of deformation temperatures from 164K to 281K at different initial structural-phase states, the constancy of γ_{SIS} and γ_{rp} is observed. With an increase in T_{def} above 299 K, a decrease in the value of the summary inelastic strain, γ_{SIS}, and a sharp increase in the plastic strain γ_{rp} are observed. This is due to the convergence of the magnitude of the yield stress with the magnitude of the martensitic shear stress.

Acknowledgments

The research was funded by Government research assignment for ISPMS SB RAS (project FWRW-2021-0004) and a grant of the President of the Russian Federation No. MK-1057.2020.8.

References

[1] Otsuka K, Re X 2005 Physical metallurgy of Ti-Ni-based shape memory alloys *Progress in Materials Science* **50** 511-678.

[2] Jani J M, Leary M, Subic A, Gibson M A 2014 A review of shape memory alloy research, applications and opportunities *Materials & Design* **56** 1078-1113.

[3] Geetha M, Dhanalakshmi K 2020 Structural Design and Realization of Electromechanical Logic Elements Using Shape Memory Alloy Wire Actuator *Physical Mesomechanics* **23** 446-456.

[4] Prokoshkin S D, Korotitskiy A V, Brailovski V, Inaekyan K E, Dubinskiy S M 2011 Crystal lattice of martensite and the reverse of recoverable strain of thermally and thermomechanically treated

Ti-Nishape-memory alloys *Physics of Metals and Metallography* **112** 170-187.

[5] Ryklina E P, Prokoshkin S D, Chernavina A A 2013 Peculiarities of implementation of abnormal high shape memory effects in thermomechanically treated Ti-No alloys *Inorganic Materials: Applied Research* **4** 348-355.

[6] Belyaev S, Rubanik V, Resnina N, Rubanik V, Lomakin I 2014 Functional properties of '$Ti_{50}Ni_{50}$-$Ti_{49.3}Ni_{50.7}$' shape memory composite produced by explosion welding *Smart Materials and Structures* **23** 085029.

[7] Ryklina E, Korotitskiy A, Khmelevskaya I, Prokoshkin S, Polyakova K, Kolobova A, Soutorine M, Chernov A 2017 Control pf phase transformation and microstucture for optimum of one-way and two-way shape memory effects in removable surgical clips *Materials & Design* **136** 174-184.

[8] Lotkov A I, Grishkov V N, Zhapova D Yu, Gusarenko A A, Timkin V N 2018 Impact of Plastic Straining in the Martensitic State on the Development of the Superelasticity and Shape Memory Effects in Titanium-Nickelide-Based Alloys *Technical Physics Letters* **44** 995-998.

[9] Zhapova D, Grishkov V, Lotkov A, Timkin V, Gusarenko A, Rodionov I 2021 Behavior of Inelastic and Plastic Strains in Coarse-Grained $Ti_{49.3}Ni_{50.7}$(at%)Alloy Deformed in B2 States *Metals* **11** 741.

[10] Zhapova D, Lotkov A, Grishkov V, Timkin V, Gusarenko A 2019 Temperature depemdence of inelastic recoverable strain and degree of shape recovery in coarse-grained samples of $Ti_{49.3}Ni_{50.7}$(at%) in the absence of plastic deformation *AIP Conference Proceedings* **2167** 020399.

IOP Conf. Series: Materials Science and Engineering 1213 (2022) 012005 doi:10.1088/1757-899X/1213/1/012005

The strain variation during the isothermal B2→B19' martensitic transformation in the quenched or annealed $Ni_{51}Ti_{49}$ alloy

A Ivanov[1], A Gabrielian[1], S Belyaev[1], N Resnina[1], V Andreev[2,3]

[1]Saint Petersburg State University, 7/9 Universitetskaya Quay, Saint Petersburg, 199034, Russia

[2]MATEK-SMA Ltd., 137, 2a Karier St, Moscow, 117449, Russia

[3]Baikov Institute of Metallurgy and Materials Science, RAS, 49 Leninskii Pr., Moscow, 119991, Russia

Email: ileavenovasky@gmail.com

Abstract. The strain variation during the isothermal holding under constant stress was studied in the quenched or annealed $Ni_{51}Ti_{49}$ alloy samples. The isothermal strain variation was found in both samples and this strain was completely recovered on subsequent unloading and heating. This allowed to conclude that the strain variation on holding was caused by the isothermal martensitic transformation. It was found that the maximum value of isothermal strain depended on the alloy heat treatment. This value was equal to 0.5 % in annealed sample and it was equal to 6 % in quenched sample. It was assumed that the formation of the Ni_4Ti_3 phase during annealing led to a decrease in concentration of substitutional Ni atoms in NiTi phase that were responsible for the isothermal transformation. As a result, the less volume fraction of the martensite formed during holding that supresses the strain variation in annealed samples.

1. Introduction

The realization of the forward B2 → B19' thermoelastic martensitic transformation under isothermal conditions was found in NiTi-based alloys with non-stoichiometric composition [1-5]. The volume fraction of the isothermal martensitic depended on the position of the holding temperature compared to the M_s temperature and on the chemical composition of the alloy [4-5]. For example, during holding at the M_s temperature, 45 % of martensite appeared isothermally in the quenched $Ni_{51}Ti_{49}$ alloy and 80 % – in the $Ti_{40.7}Hf_{9.5}Ni_{44.8}Cu_5$ alloy [4-5]. In a quenched Ni-rich NiTi alloy, excess Ni atoms replace Ti atoms in the Ti sub-lattice and become the substitutional defects. In the quaternary Ti-Hf-Ni-Cu alloy, Hf and Cu atoms replace the Ti and Ni atoms in their sub-lattices, and act as substitutional defects [6]. It was assumed that the substitutional defects prevented from the shift of atoms needed for the transformation of the B2 phase to the B19' phase. As a result, the strain nanodomains formed instead of the B19' phase. During holding the migration of the substitutional defects took place the allowed to complete the shift of atoms and the B19' phase appeared. As the concentration of the doping atoms in $Ti_{40.7}Hf_{9.5}Ni_{44.8}Cu_5$ alloy is higher than the excess of Ni atoms in the $Ni_{51}Ti_{49}$ alloy, hence it may be concluded that a concentration of substitution defect in the $Ti_{40.7}Hf_{9.5}Ni_{44.8}Cu_5$ alloy is larger than in quenched $Ni_{51}Ti_{49}$ alloy. Thus, one may conclude that a decrease in substitution defects decreases the

volume fractions of isothermal martensite. If the concentration of substitutional defect is zero, hence no isothermal transformation occurs on holding that has been confirmed in [7].
Previously, it was found that the isothermal transformation was accompanied by the recoverable strain variation if holding was carried out under a stress [8-9]. If a decrease in the substitutional defects reduces the volume fraction of the isothermal martensite, hence it should affect the isothermal strain variation. The concentration of substitutional defects may be changed by annealing. In quenched Ni-rich NiTi alloy, all excess Ni atoms act as substitutional defects [6]. Annealing of Ni-rich NiTi alloy leads to the Ni_4Ti_3 precipitates formation that decreases the Ni concentration in the NiTi phase [6]. Other words, annealing decreases the concentration of substitutional defects in Ni-rich NiTi alloy. Thus, the quenched and annealed Ni-rich NiTi samples are characterized by different concentration of the substitutional defects that should affect the isothermal strain variation. The aim of the present work was to study this assumption.

2. Materials and Methods

Wire samples of $Ni_{51}Ti_{49}$ alloy (Matek-SMA Ltd., Moscow, Russia) with a diameter of 1.5 mm were water quenched from 850 °C for 15 minutes. Some quenched samples were subjected to 100 thermal cycles in a temperature range of -196 °C to 100 °C to stabilize the transformation temperatures. Other samples were annealed at a temperature of 450 °C for 2 hours. Calorimetric study shows that quenched samples undergo B2 ↔ B19' transformations at temperatures of M_s = -43 °C, M_f = -63 °C, A_s = -24 °C, A_f = -9 °C (after 100 thermal cycles). Annealed samples undergo the B2 → R → B19' transformations on cooling and the B19' → B2 transformation on heating at temperatures: R_s = 36 °C, R_f = 27 °C, M_s = -20 °C, M_f = -30 °C, A_s = 32 °C, A_f = 40 °C.
To study the strain variation during the isothermal martensitic transformation, the following procedure was used. The sample was cooled to the holding temperature T*, held for 40 minutes to stabilize the temperature, loaded to 300 MPa, held for 60 minutes, unloaded and heated through a temperature range of the reverse transformation. A stress of 300 MPa was chosen due to both alloys underwent the B2 ↔ B19' transformation on cooling/heating under this stress or loading to 300 MPa and unloading. The holding temperatures were chosen to be close to the start temperature of the forward B2 → B19' transformation under a stress of 300 MPa (M_s^{300}). Quenched samples were held at temperatures close to M_s^{300} = -8 °C and annealed samples were held at temperatures close to M_s^{300} = 48 °C.

3. Results and Discussions

Figure 1 shows the σ(ε) curves obtained on loading-unloading (red line) and loading-holding-unloading (red line) of annealed and quenched samples. In annealed sample, strain increased to 2.5 % on loading to 300 MPa and completely recovers on unloading (Figure 1a). During holding, strain additionally rises by 3 % and fully recovered after unloading and subsequent heating. In quenched sample, loading to 300 MPa leads to increase in strain to 0.8 % that completely recovers on unloading. Holding of the sample after unloading is accompanied by an increase in strain to 6.8 % (Figure 1b). This strain recovers on unloading but some irreversible strain of 0.7 % is observed. Thus, the results of the study show that in both quenched and annealed samples, the isothermal strain variation takes place on holding under 300 MPa and this strain partially or completely recovers on unloading. This means that the strain variation on holding is caused by reversible deformation mechanism. As the strain variation occurs on holding even at the M_s^{300} temperature, hence the martensite reorientation can not be responsible for this phenomenon. In this case, the reason for the isothermal strain variation is the isothermal formation of the oriented martensite. On unloading the martensite is unstable and transforms to austenite that is accompanied by the strain recovery.

IOP Conf. Series: Materials Science and Engineering 1213 (2022) 012005 doi:10.1088/1757-899X/1213/1/012005

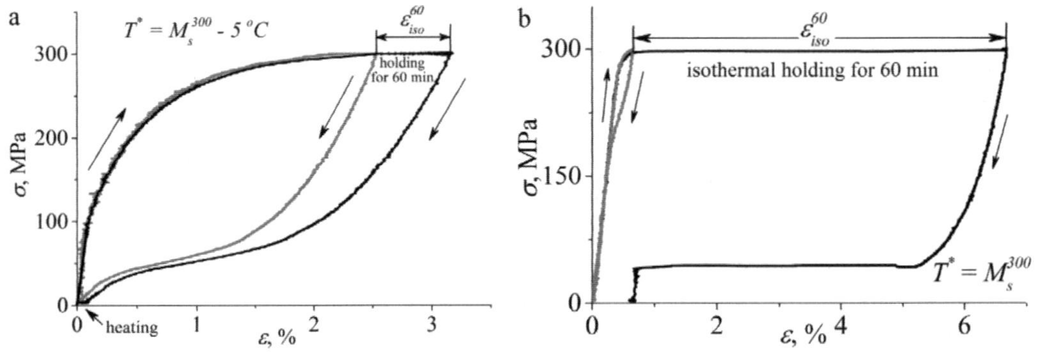

Figure 1. The stress-strain curves obtained on loading and unloading (red line) and on loading – holding – unloading (black line) of the annealed (a) and quenched (b) $Ni_{51}Ti_{49}$ samples.

To study the influence of the position of the holding temperature compared to the M_s^{300} temperature on the isothermal strain variation, the holding under a stress of 300 MPa was carried out at various temperatures (Figure 2). It was observed that in both samples, the strain increased up to saturation which value depended on the holding temperature and heat treatment of the sample.

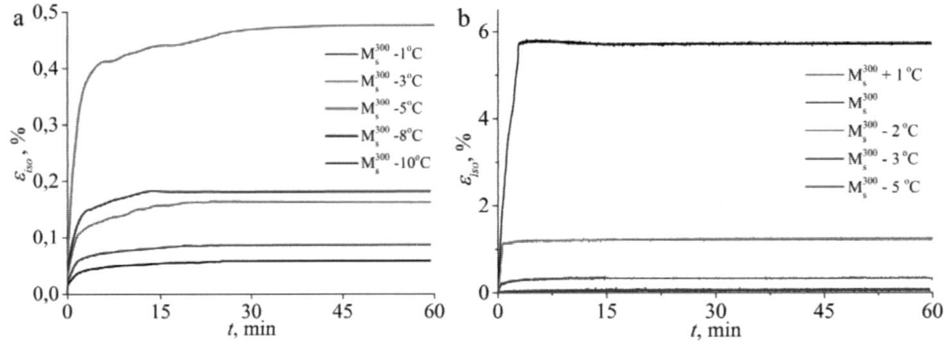

Figure 2. The strain variation on holing under a stress of 300 MPa at different holding temperatures in the annealed (a) and quenched (b) $Ni_{51}Ti_{49}$ alloys.

Figure 3 shows, the dependence of maximum isothermal strain (ε_{iso}^{60}) accumulated in the sample during holding for 60 min on the difference *(ΔT)* between the holding temperature (*T**) and the M_s^{300} temperature. It was found that the dependence of the isothermal strain on *ΔT* was non-monotonic for both alloys. The position of the maximum was different for the quenched and annealed samples: in quenched sample the maximum was observed on holding at the M_s^{300} temperature whereas, in annealed sample maximum was found at M_s^{300} - 5 °C temperature. Moreover, the maximum value of the isothermal strain in quenched sample was 10 times larger than in annealed sample.

Therefore, the results of the study show that the isothermal strain variation on holding under 300 MPa in quenched $Ni_{51}Ti_{49}$ alloy was 10 times larger than in annealed $Ni_{51}Ti_{49}$ alloy. This confirms the assumption that a decrease in substitution defect concentration partially suppresses the isothermal martensitic transformation that reduces the isothermal strain variation on holding under a stress. Thus, to observed the isothermal strain variation in the Ni-rich NiTi alloys, the quenched samples should be used otherwise, to suppress the isothermal strain variation, the samples should be annealed.

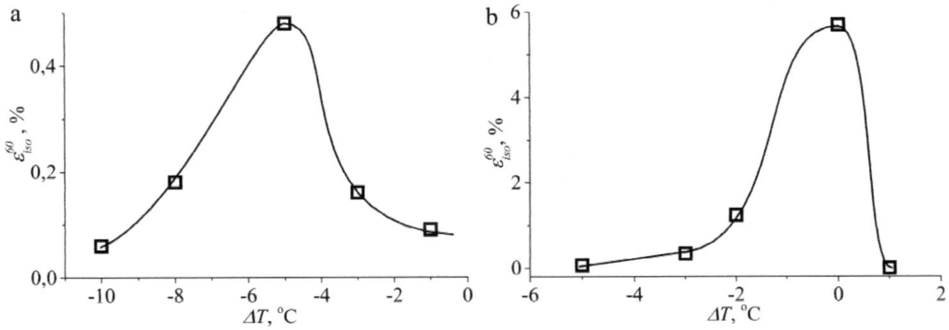

Figure 3. The dependences of the isothermal strain on the position of the holding temperature compared to the M_s^{300} temperature found in the annealed (a) or quenched (b) $Ni_{51}Ti_{49}$ samples

4. Conclusions

1. The strain variation was observed during holding at 300 MPa in annealed or quenched $Ni_{51}Ti_{49}$ samples after active deformation at temperatures close to the M_s^{300} temperature.
2. The maximum isothermal strain variation depends on the position of the holding temperature compared to the M_s temperature under a stress of 300 MPa.
3. The maximum isothermal strain variation depends on the heat treatment of the Ni-rich NiTi alloys which affects the concentration of the substitutional defects. In quenched samples, all excess of Ni atoms act as substitutional defects. In annealed Ni-rich NiTi alloys, the concentration of the substitutional defects is less than in quenched alloys due to the formation of the Ni_4Ti_3 precipitates. The less the concentration of substitutional defects the less the volume fraction of the isothermal martensite and as result, the less the isothermal strain variation.

Acknowledgments
This work was supported by Russian Science Foundation (grant number 18-19-00226-P).

References

[1] Kustov S, Salas D, Cesari E, Santamarta R, Van Humbeeck J 2012 Acta Mater. 60 2578-2592
[2] Fukuda T, Yoshida S, Kakeshita T 2013 Scripta Mater. 68 984-987
[3] Fukuda T, Kawamura T, Kakeshita T 2016 J Alloys and Compd. 683 481-484
[4] Resnina N, Belyaev S, Demidova E, Ivanov A, Andreev V 2018 Mater. Lett. 228 348-350
[5] Demidova E, Belyaev S, Resnina N, Shelyakov A 2020 J Thermal Analysis Calorimetry 139 2965-2970
[6] Otsuka K, Ren X 2005 Progr. Mater. Sci. 50 511-678
[7] Otsuka K, Ren X, Takeda T 2001 Scripta Mater. 45 145-152
[8] Ivanov A, Belyaev S, Resnina N, Andreev V Sensors and Actuators A: 297 111543
[9] Demidova E, Belyaev S, Resnina N, Shelyakov A 2019 Mater. Lett. 254 266-268

IOP Conf. Series: Materials Science and Engineering 1213 (2022) 012006 doi:10.1088/1757-899X/1213/1/012006

Nucleation centers for martensite with habitus {110} in the shape memory alloys

M P Kashchenko[1,2*] **N M Kashchenko** [1] **and V G Chashchina**[1,2]

[1] Ural Federal University, 19 Mira str., Yekaterinburg , 620002, Russia
[2] Ural State Forestry University, 37 Sibirskiy trakt str., Yekaterinburg, 620100, Russia

* Corresponding author: mpk46@mail.ru

Abstract. The dynamic theory of martensitic transformations (MT) considers the formation of habit planes of martensite crystals as a consequence of the propagation of a controlling wave process (CWP). The general ideology makes it possible, by comparing the observed habits with calculations of the elastic fields of defects (as a rule, dislocations) , to identify nucleation centers. In a number of cases (In-Tl alloys, $Ni_{50}Mn_{50}$ alloys, Heusler alloys ...) under MT in the shape memory alloys, {110} habits are observed (in the basis of the initial cubic phase), which often have a fine twin structure with twin boundaries of the same type. This highly symmetric structure is described by the CWP containing longitudinal waves (both relatively long-wavelength ℓ and short-wavelength s) propagating along the 4-order symmetry axes. In this paper, it is shown that such habits are associated with rectilinear segments of dislocation loops with directions Λ along <001> and Burgers vectors along <010> (or <110>) orthogonal to Λ, both for sliding and for prismatic loops. The tetragonality, the relative volume change during the MT, and the dependence of the start temperature M_s on changes in the concentration of alloy components are also briefly discussed.

1. Introduction

Martensitic transformations (MTs) provide a very common way of realizing polymorphic transformations in crystalline solids. The main feature of MTs is their cooperative nature, which indicates the existence of a lattice restructuring control mechanism. Understanding the essence of this mechanism is the key to explaining all the observed features of the MT. In particular, the set of morphological features is the main characteristic of MT. Macroscopic features includes habit plane orientation, interfacial orientation of crystal lattices, and macroscopic shear. All these features are unambiguously correlated with each other.

This directly indicates that the morphological features are a consequence of the action of a single mechanism that controls crystal growth. But then it is natural to believe that a set of morphological features can be used to reconstruct (although not completely) a dynamic (evolving in time and space) picture of transformation. At present, the solution of the direct problem of identifying the mechanism for controlling the growth of a martensite crystal and the inverse problem of restoring the transformation dynamics by morphological characteristics has been performed for the fcc-bcc (or bct) transformations in iron alloys, as well as for the bcc-hcp MT in titanium [1-3].

The indicated MT variants belong to pronounced first-order transitions. Moreover, experimental studies reporting a supersonic growth rate of martensite crystals directly indicate the existence of a control wave process (CWP). The consistent development of ideas about the wave mechanism of crystal growth control made it possible to fundamentally supplement the scenarios of heterogeneous nucleation for transformations of the first kind in a solid.

The concept of an initial excited state (IES) has replaced the concept of the existence of quasi-equilibrium nuclei as regions of the initial phase separated by defects. IES arise in elastic dislocation fields at temperatures M_s below the phase equilibrium temperature T_0 as a result of a jump-like transition of atoms to new equilibrium positions. The region of localization of the IES has the shape of an elongated rectangular parallelepiped, the orientation of the edges of which is specified by the

IOP Conf. Series: Materials Science and Engineering 1213 (2022) 012006 doi:10.1088/1757-899X/1213/1/012006

eigenvectors of the strain tensor corresponding to the elastic field of a dislocation playing the role of a nucleation center (NC) for a martensite crystal. The type of excitation is an oscillatory process. The initial vibration amplitudes depend on the degree of deviation of the temperature M_s from T_0. It is the IES that generates the wave process that overcomes the threshold deformation. As shown in [4-6], the CWP can inherit the deformation in the area of the IES. But, knowing the nucleation centers, this deformation can be calculated for known elastic moduli of the initial phase.

As a result, there is a fairly transparent algorithm for comparing morphological features with the calculation of elastic fields of NC. The simplest for the calculation is to determine the orientations of the habit planes, which are compared to the motion of the line of intersection of the wave beams included in the CWP. Moreover, relatively long-wavelength ℓ-displacements of atoms are responsible for the formation of habits, and relatively short-wavelength s-displacements are responsible for the formation of twin boundaries [1].

In the case of shape memory alloys (SMA) the level of threshold deformations is lower than in the case of γ-α MT. However, as the analysis shows [7], the observed morphological features of martensite crystals in alloys based on titanium nickelide can be unambiguously compared by the NCs for all variants of martensitic transformations. This result testifies to the possibility of using the concept of heterogeneous nucleation.

Note that in the case of the existence of the temperature T_c of the absolute loss of stability of the initial phase, when T_c is reached, the homogeneous nucleation is inevitable. However, NCs initiate the start of the formation of martensite crystals at temperatures $M_s > T_c$. In real (even strongly annealed) crystals, dislocations are always present; therefore, the onset of an MT is due to heterogeneous nucleation. This fundamental conclusion is also confirmed with the framework of the molecular dynamics method [8].

An important difference between MT in SMA is the possibility of significant lateral crystal growth, which is associated with a high degree of coherence of phase lattices at the interface and a high elastic limit of the initial phase. This ensures a high degree of transformation reversibility and its thermoelastic character. Within the framework of the dynamic theory, the main attention is paid to the formation of the central part of crystals, but the mechanism of lateral growth can also be described with somewhat less certainty.

In a number of cases (In-Tl, Ni50Mn50 alloys, Heusler alloys ...) under MT in SMA, {110} habits are observed (in the crystallographic basis of the initial cubic phase), which often have a fine twin structure with orientations of twin boundaries from the same set {110}. This highly symmetric structure is described by a CWP containing purely longitudinal ℓ- and *s*-waves propagating along the <001> symmetry axes. This option corresponds to the simplest version of inheriting the characteristics of the elastic field of the NC, in which the directions of the wave normals of the two ℓ-longitudinal waves responsible for describing the habit coincide with the orientations of the eigenvectors of the deformation tensor of the elastic field of the NC. It is the high symmetry of the habit planes that made it possible earlier in [9, 10] to immediately use purely longitudinal waves in the analysis.

The main goal of this work is to show that there are NCs that make it possible, within the framework of the developed dynamic model, to describe the observed {110} habits, as well as to briefly discuss the expected tetragonality of martensite, the specific change in volume upon MT, and the dependence of M_s on the chemical composition.

2. Identification of NC in the case of crystals with {110} habit planes

It is natural to choose a dislocation loop with a rectilinear segment Λ_1 along the fourth-order axis, for example, along [001] with the Burgers vector $\boldsymbol{b}$ of the edge (with respect to Λ_1) orientation, as the expected variant of NC. In particular, for the initial bcc phase with the B2type ordering, this can be one of the vectors $\boldsymbol{b}_1 \parallel [100]$ or $\boldsymbol{b}_2 \parallel [010]$.In addition to the indicated Burgers vectors, $\boldsymbol{b}_3 \parallel [110]$ and $\boldsymbol{b}_4 \parallel [-110]$ were also considered. Assuming the loops to be rectangular, the plane of the loop was varied by choosing the orientation of the second segment Λ_2.Cases of sliding and prismatic loops were considered. The set (in units of 10^{11} dyn / cm^2) C_{44} = 10.3, C_L = 25, C' = 0.445, taken from [11], was used as the elastic moduli, which reflects the small value of the modulus C' typical for the SMA.

The calculation of the elastic field of the NC was carried out in a cylindrical coordinate system (with the z axis along Λ_1). The origin of coordinates was chosen in the middle of the segment Λ_1, the angle θ was measured from the plane of the loop. All linear dimensions were specified when choosing the parameter of the initial cubic lattice as a unit of length. The value L_l = 7000 was used as the typical size of the length L_l of the segment Λ_1. The L_2 length of the Λ_2 segment was chosen to be sufficiently large ($L_2 = 10^4$), and the distance R = 300 from the origin of coordinates was noticeably smaller than the L_2 loop width, so that the Λ_1 segment played the main role in the deformation.

The angular dependences of the eigenvectors $\xi_{1,2}$ (the vector ξ_3 is directed along the z axis) and the eigenvalues $\varepsilon_{1,2}$ ($\varepsilon_3 \approx 0$) of the strain tensor, the magnitude and direction of the shear vectors **S**, and the relative volume change δ were calculated. For illustration, Figs. 1 and 2 show the results of calculating ε_1 (θ) > 0 (tension), ε_2 (θ) < 0 (compression), and δ (θ) on a scale that is convenient for perception.

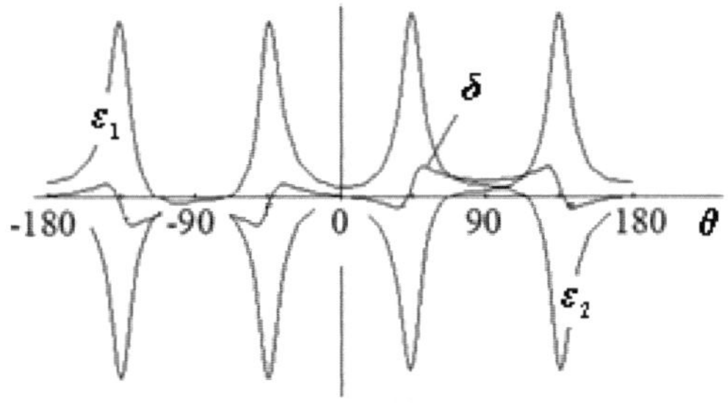

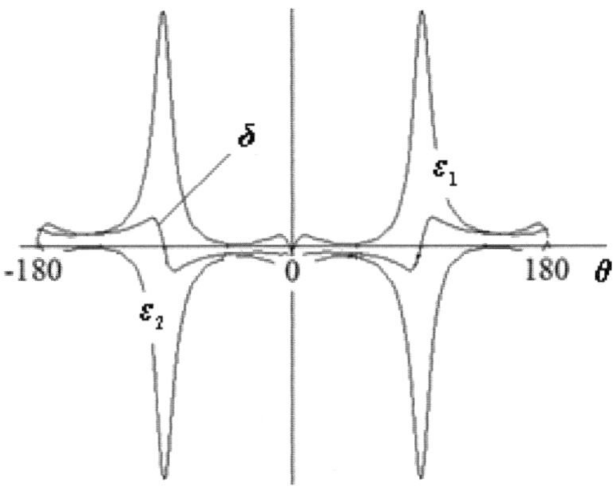

Figure 1. Angular dependences of deformations $\varepsilon_{1,2}$ and relative change in volume δ in the case of the Burgers vectors $\boldsymbol{b}_2 \parallel$ [010] in (a) for a dislocation in the cubic slip plane (001) and $\boldsymbol{b}_3 \parallel$ [110] in (b) for a prismatic dislocation with segments of a rectangular loop $\Lambda_1 \parallel$ [001], $\Lambda_2 \parallel$ [-110].

It is characteristic that the regions of extrema of the dependences $\varepsilon_{1,2}$ and shear (at $|\theta| = 45^0$, 135^0 in (a) and at $|\theta| = 90^0$ in (b)) correspond to the orientations of the strain eigenvectors $\xi_{1,2}$, which are almost collinear to the directions [100] and [010]. In the vicinity of these angles, the volume δ changes sign. Thus, rectilinear <001> segments of dislocation loops with the edge orientation of the Burgers vector with respect to them act as NCs for the formation of crystals with {110} habits.

It is interesting to note that the origin of the NC can be associated with the contact interaction of dislocations with intersecting slip planes. For example, if we consider the cubic slip planes as the initial ones, then the simplest barrier arising in the contact interaction of a pair of dislocations (100) [010] and (010) [100] is characterized by the line $\Lambda \parallel$ [001] and the Burgers vector along [110].On the contrary, taking two slip systems of a disordered bcc lattice (110) [-111] and (1-10) [11-1], we have the line of intersection $\Lambda \parallel$ [001] and the Burgers vector [010]. It is clear that in the case of contact interaction of dislocations, the resulting barriers are relatively stable. Then it is easy to understand why there are stable centers in the vicinity of which the start of the formation of martensite crystals during thermal cycling is repeatedly reproduced.

3. Simplification of the description of the dependence on deformation of the degree of tetragonality and the relative change in volume for martensite crystals with {110} habits

IOP Conf. Series: Materials Science and Engineering 1213 (2022) 012006 doi:10.1088/1757-899X/1213/1/012006

In the case of crystals with (110) habits, it is easy to associate the tetragonality of t martensite to one of the main values of the Bain strain tensor. Although we discussed this issue in [9, 10], it is useful for the convenience of the reader to recall the arguments presented.

In the dynamic theory of MT [1-3], the kinematic description of the habit in the threshold mode corresponds to the deformation description. From the requirement that these descriptions coincide, it follows that the ratios of the moduli of compression deformations $\varepsilon_2 < 0$ and tension $\varepsilon_1 > 0$ in the threshold mode coincide with the ratios of the squares of the wave velocities. However, when longitudinal waves propagate along orthogonal symmetry axes from the <001> family, the ratio of their velocities is 1, therefore

$$\varepsilon_1/|\varepsilon_2| = 1. \tag{1}$$

Then it is natural to assume that the unstable austenite lattice, transforming to new stable positions of atoms, retains the ratio of deformations in the range from threshold to final values.

In addition, we assume that the first fast stage of the transformation is associated with short-wavelength s-deformations (for a time on the order of T_s of the s-oscillation period), and already at this stage the final (Bain ε_{2B}) compression deformation in the [100] direction $\varepsilon_{2s} = \varepsilon_{2B} < 0$ is achieved. Since the ratio of velocities for s –waves is also equal to 1, the tensile strain along [010] at the first stage reaches the value $\varepsilon_{1s} = -\varepsilon_{2s} = -\varepsilon_{2B}$.

Since at the second stage (for a time of the order of T_ℓ of the ℓ-oscillation period), compression and tension are also carried out by waves with the same velocities, the deformation contributions $\varepsilon_{2\ell} = -\varepsilon_{1\ell}$. If the bcc - fcc MT is considered, then it follows from the symmetry of the fcc lattice of martensite that the compression deformation $\varepsilon_{2\ell}$ along the [001] axis is equal to the deformation $\varepsilon_{2s} = \varepsilon_{2B}$. Consequently, $\varepsilon_{1\ell} = -\varepsilon_{2B}$, that is, at the second stage, the contribution of the ℓ-wave to the main Bane tensile deformation is equal to the contribution from the s-wave at the first stage. We further take into account that, in view of the inequality $T_\ell \gg T_s$, the ℓ-tension already acts on the stretched lattice. As a result, the linear size a (lattice parameter of bcc austenite) in the [010] direction increases by $(1 + |\varepsilon_{2B}|)^2$ times, and the transverse size decreases by $(1 - |\varepsilon_{2B}|)$ times.

Hence, to determine t, one should take the ratio of the size of the edge of the cube of the initial unit cell of the bcc lattice after tensile deformation to the size of the diagonal of the face after compression deformation:

$$t_\gamma = (1+|\varepsilon_{2B}|)^2/[(1 - |\varepsilon_{2B}|)\sqrt{2}\,]. \tag{2}$$

According to (2), the value $t_\gamma = 1$ corresponds to $|\varepsilon^*_{2B}| \approx 0.1173$, $t_\gamma < 1$ at $|\varepsilon_{2B}| < |\varepsilon^*_{2B}|$ and $t_\gamma > 1$ at $|\varepsilon_{2B}| > |\varepsilon^*_{2B}|$. Under the same assumptions, we can easily find the relative change in the specific (per atom) volume δ:

$$\delta = -\varepsilon^2_{2B}(2 - \varepsilon^2_{2B}) \approx -2\varepsilon^2_{2B} < 0. \tag{3}$$

This is illustrated by the data in table 1.

Table 1. Values of t_γ and δ depending on $-\varepsilon_{2B}$

$-\varepsilon_{2B}$	t_γ	$-\delta$
0	$1/\sqrt{2}$	0
0.11	0.979	0.024
$-\varepsilon^*_{2B}$	1	0.027
0.15	1.100	0.04
0. 2	1.273	0.078

That is, at values of $\varepsilon_{1B} \sim 0.1$, the volume effect $\delta \sim -0.02$, but at $\varepsilon_{1B} \sim 0.2$, it is almost four times larger (see Table 1). We recall [12] that, in contrast to the ℓ-waves initiated by IES in the elastic field of a dislocation, s-waves are associated with geometrically similar (IES)$_s$, which have a significantly smaller transverse size and arise spontaneously.

If the CWP contains only a pair of ℓ -waves, then in the ξ_3 direction due to electronic correlations there should be compression (at bcc-fcc MT), restoring the lattice symmetry. In this case, the increase in tetragonality with an increase in $\varepsilon_{1B} = \varepsilon_{2B}$ would occur more slowly, according to the relation

$$t_\gamma = (1+|\varepsilon_{2B}|)/[(1 - |\varepsilon_{2B}|)\sqrt{2}\,], \tag{4}$$

and the volume effect for the same $|\varepsilon_{2B}|$ would decrease by almost 2 times

$$\delta = (1+|\varepsilon_{2B}|)[(1 - |\varepsilon_{2B}|) - 1 = -\varepsilon^2_{2B} < 0. \tag{5}$$

The value $t_\gamma = 1$ then corresponds to the value $|\varepsilon^*_{2B}| = [3 - 2\sqrt{2}\,] \approx 0.171573$. Other values are shown for illustration in table 2.

Table 2. Values of t_γ and δ depending on $-\varepsilon_{2B}$ without the contribution of s-waves

$-\varepsilon_{2B}$	t_γ	$-\delta$
0	$1/\sqrt{2}$	0
0.11	0.882	0.012
$-\varepsilon^*_{2B}$	1	0.029
0.2	1.061	0.04
0. 27	1.230	0.073

Comparison of the data in tables 1 and 2 shows that at t_γ near 0.98, as well as near 1.2, the relative volume changes are close, although the deformation values themselves are noticeably larger in the absence of the *s*-waves contribution.

4. Discussion of results

It should be expected that the phase with a large tetragonality $t_\gamma \approx 1.2$ will have a lower resistance to thermal cycling as compared with the variant $t_\gamma \approx 0.98$. Indeed, during the formation of crystals with {110} habits, the conditions of deformation with an invariant plane are satisfied. Therefore, the deformation equal to δ is realized in the direction orthogonal to the habit plane. This means that for large δ and t_γ, the probability of reversible thermoelasticity should decrease. Of course, the value of the yield point should also be taken into account. If it is not exceeded, then the variant with large δ and t_γ can remain stable during thermal cycling.

It is clear that variants of martensite with a large value of δ should mainly arise at small (sharp) angles of habit with the outer surface, for example, of film samples or foil, since in this case the resistance to volume change will be relatively small.

As already noted, within the framework of the dynamic theory, the main attention is paid to the formation of the central part of the crystals, which specifies the orientation of the habit. It should be borne in mind that at small values of δ, the direction of the macroshear lies in the habit plane (-110) and is almost collinear [110]. This conclusion is valid for both the Burgers vector [010] and [110]. This means that the end face of the resulting crystal should play the role of a NC with a $\boldsymbol{\Lambda}_1$ segment and a Burgers vector [110]. In essence, a relay-race mechanism for reproducing an elastic field of the same type as the field of one of the above-considered NCs is realized. And such reproduction will occur during the formation of any martensite crystal with a habit from the {110} family.

As one of the mechanisms of lateral growth, one can consider the variant of the propagation of the softest transverse mode in the direction of the normal to the habit plane. This mode is capable of both increasing and partially compensating for the amount of shear in the direction orthogonal to the normal to the habit plane. It is also possible to increase the crystal thickness due to the path of pairs of s-waves excited by spontaneously arising $(IES)_s$ in the regions of the initial phase adjacent to the habit.

The results are shown in table 1, are valid for sufficiently large grain sizes of polycrystalline samples, when the ranges of the wave vectors of the ℓ- and s-waves differ in order of magnitude, but the dispersion of the s-waves can still be ignored. However, for the nanocrystalline state, as in the consideration of degenerate twin structures [13, 14], it is necessary to take into account the dispersion of s-waves.

It is of interest to extend the conclusions of the dynamic theory of γ-α MT in iron alloys, concerning the dependence of the temperature M_s on the chemical composition of the alloy and the grain size, to SMA materials.

In the case of disordered alloys with a dominant iron content, these questions were resolved under the following assumptions. 1. The main role in the generation (amplification) of waves initiated by the IES is played by 3d electrons, the energies of which belong to the region of the peak of the density of states associated mainly with iron atoms. 2. Active (at the growth stage) pairs of inversely populated states of electrons in the quasi-pulse space are separated by nonequipotential *S*-surfaces, on which the nonequilibrium addition to the Fermi distribution function vanishes. 3. The energy of electrons that are important for the generation of elastic waves should be spaced from the Fermi level (chemical potential μ) in the range up to 0.2 eV. 4. Population of 3*d*–electronic states is carried out with the participation of strongly damped states of the *s*–band, through the redistribution of *d*–electrons in the processes of *d*–*s*–*d* scattering. In this case, the damping of *s*–electrons Γ_s plays a role equivalent to temperature for *d*–electrons. So the addition of an impurity leading to an increase in Γ_s should be accompanied by a decrease in the temperature contribution to the electron distribution and, at the same time, by a decrease in the temperature M_s. Obviously, with this approach, the role of alloy components additional to iron is reduced mainly to electrical inhomogeneities, on which s-electrons are scattered. The influence of changes in the concentration of components on the value of ε_d - μ (in the spirit of the rigid zone ideology) is neglected, since it is assumed that each of the elements forms its own d-zone.

There is no doubt that, for SMA, the features of the MT are associated with the ensemble of electrons belonging to the region of increased density of states in the acceptable energy range $\varepsilon_d - \mu$ ~ 0.1 eV. However, as a rule, we are talking about alloys in which the starting chemical composition contributes to the formation of an ordered superstructure. In this case, the peak of states necessarily includes the partial contributions of the alloy components. A deviation from stoichiometric proportions with a change in the starting composition can then be accompanied by a broadening of the peak of the density of states and a change in ε_d - μ. Most likely, this is precisely why the rigid band ideology leads to satisfactory agreement with the observed change in M_s in Heusler alloys, when the dependence of M_s on the component concentrations is reduced to the dependence on the concentration of the total number of $d + s$ electrons per atom [15].

5. Conclusions

The nucleation centers for martensite crystals with {110} habits are segments of dislocation loops along the <001> axes with edge (with respect to the segments) orientations of the Burgers vectors along <010> or <110> directions.

There are reasons to believe that a number of results of the dynamic theory of MT, developed for MT in iron-based alloys, can be extended to MT in alloys with shape memory effect.

The most promising directions of application and development of the dynamic theory of MT to alloys with a thermoelastic transformation are related to the description of the mechanism of lateral growth of a martensite crystal and verification of the applicability of the formula for the critical grain size of the initial phase [16, 17] for the implementation of the martensitic transformation.

Acknowledgments

The authors express their gratitude to the Ministry of Science and Higher Education of Russia for their support in the execution of state assignment No. 075-00243-20-01 dated 26.08.2020 within the framework of the theme FEUG-2020-0013 “Environmental aspects of rational nature management”.

References

[1] Kashchenko M P and Chashchina V G 2011 *Phys.Usp.* **54** 331.
[2] Kashchenko M P and Chashchina V G 2008 *Phys. Met. Metallogr.***105** 537
[3] Kashchenko M P and Chashchina V G 2008 *Phys. Met. Metallogr.***106** 14
[4] Kashchenko M P, Chashchina V G and Latypov I F 2015 *Russian Phys. Journal* **58** 79
[5] Kashchenko M P, Chashchina V G and Latypov I 2016 *Russian Physics Journal* **59** 750
[6] Kashchenko M P and Chashchina V G 2017 *Metal Science and Heat Treatment* **59** 410
[7] Kashchenko M P and Chashchina V G 2015 *Materials Science Foundations: Proceedings* **81-82** 3
[8] Dmitriev S V, Kashchenko M P, Baimova Ju A, Babicheva R I, Gunderov D V and Pushin V G 2017 *Lett. Mater.* **7** 442
[9] Kashchenko M P, Kashchenko N M and Chashchina V G 2018 *Lett. Mater.* **8** 429
[10] Kashchenko M P, Kashchenko N M, Chashchina V G, Belosludtseva E S, Pushin V G and Uksusnikov A H 2019 *Phys. Met. Metallogr.***120** 782
[11] Worgull J, Petti E and Trivisonno J 1996 *Phys. Rev.* B **54** 15695
[12] Kashchenko M P, Kashchenko N M and Chashchina V G 2019 *Phys. Met. Metallogr.***120** 422
[13] Kashchenko M P, Kashchenko N M and Chashchina V G 2019 *Phys. Solid State* **61** 2254
[14] Kashchenko M P, Kashchenko N M and Chashchina V G 2020 *Mechanics of Solids* **55** 6
[15] Chernenko V A 1999 *Scripta Mater.* **40** 523
[16] Kashchenko M P, Kashchenko N M, Korolev A V, Chashchina V G and Oglezneva S A 2019 *Phys. Mesomech* .**22** 203
[17] Kashchenko M P, Kashchenko N M and Chashchina V G 2021 *Phys. Met. Metallogr.* **122** 47

IOP Conf. Series: Materials Science and Engineering 1213 (2022) 012007 doi:10.1088/1757-899X/1213/1/012007

Strain variation in the $Ti_{40,7}Hf_{9,5}Ni_{41,8}Cu_8$ alloy during isothermal martensitic transformation under a constant stress

E Demidova[1], P Pchelnikov[1], S Belyaev[1], N Resnina[1] and A Shelyakov[2]

[1]Saint-Petersburg State University, 7/9 Universitetskaya nab., St. Petersburg, 199034, Russia
[2]National Research Nuclear University MEPhI (Moscow Engineering Physics Institute), 31 Kashirskoe shosse, Moscow, 115409, Russia

*Corresponding author's e-mail: lena-demi@yandex.ru

Abstract. Strain variation on holding under a constant stress was studied in the $Ti_{40,7}Hf_{9,5}Ni_{41,8}Cu_8$ alloy. It was found, that on holding under stress, the isothermal strain rose up to saturation, which value depended on holding temperature and stress. It was found that the dependencies of the isothermal strain on the holding temperature and stress were non-monotonic. This allowed to find the optimal value of stress and time at which the isothermal strain attained the maximum value of 3.2 %. It was found that the maximum isothermal strain in the $Ti_{40,7}Hf_{9,5}Ni_{41,8}Cu_8$ alloy was less than in the $Ti_{40,7}Hf_{9,5}Ni_{44,8}Cu_5$ alloy.

1. Introduction

Thermoelastic martensitic transformations in NiTi-based shape memory alloys were believed to be characterized by athermal kinetics hence, the phase transformation could not occur during isothermal holding [1,2]. However, recently it was shown, that in NiTi-based alloys with substitutional defects, the martensite phase might appear during holding at constant temperature close to M_s [3-6]. For instance, the isothermal martensitic transformation was found in the $Ti_{40,7}Hf_{9,5}Ni_{44,8}Cu_5$ alloy, where Hf and Cu atoms substituted the Ti and Ni atoms and acted as substitutional defects. The kinetics of this process was well studied and it was shown, that during isothermal holding, the martensite volume fraction increased up to saturation, which value depended on holding temperature and chemical composition of the alloy [6,7]. Moreover, in [8,9] it was shown, that the forward martensitic transformation during isothermal holding under a stress was accompanied by recoverable strain variation.

The volume fraction of the martensite that appeared on isothermal holding without stress depended on the chemical composition of the alloys. Thus, one may assume that the strain variation induced by the isothermal transformation on holding under a stress may be sensitive to the chemical composition of the alloy. To verify this assumption, the strain variation during isothermal holding of the $Ti_{40,7}Hf_{9,5}Ni_{41,8}Cu_8$ alloy under a constant stress was studied and compared to the results found previously in the $Ti_{40,7}Hf_{9,5}Ni_{44,8}Cu_5$ alloy.

2. Materials and methods

Amorphous $Ti_{40,7}Hf_{9,5}Ni_{41,8}Cu_8$ thin ribbon with a size of 15 × 1,6 × 0,04 mm was subjected to crystallization during isothermal holding at 470 °C for 30 min. The crystallized alloy underwent the B2↔B19′ transformation at temperatures of M_s = -8 °C, M_f = -14 °C, A_s = 39 °C, A_f = 48 °C. Mechanical experiments were carried out in the tensile mode using Lloyd 30K Plus testing machine, equipped with thermal chamber and video extensometer. To study the isothermal strain variation, the sample was cooled under a constant stress to holding temperature, held for 60 minutes and then heated through a temperature range of the reverse transformation (see details in [8,9]). Figure 1 demonstrates the typical strain and temperature variation on cooling, holding and heating under a stress. It is seen, that strain increases on cooling under a stress that is typical for NiTi-based alloys and caused by the formation of oriented martensite [1]. However, an additional strain variation is observed on holding under the same stress. On heating strain accumulated on cooling and holding under a stress completely recovers due to the reverse transformation. A_s complete strain recovery occurs on heating hence, the strain variation on cooling and holding is caused by the formation of oriented martensite. The strain variation on holding was studied at various holding temperatures close to M_s and under a different stress from a range of 160 MPa ÷ 400 MPa.

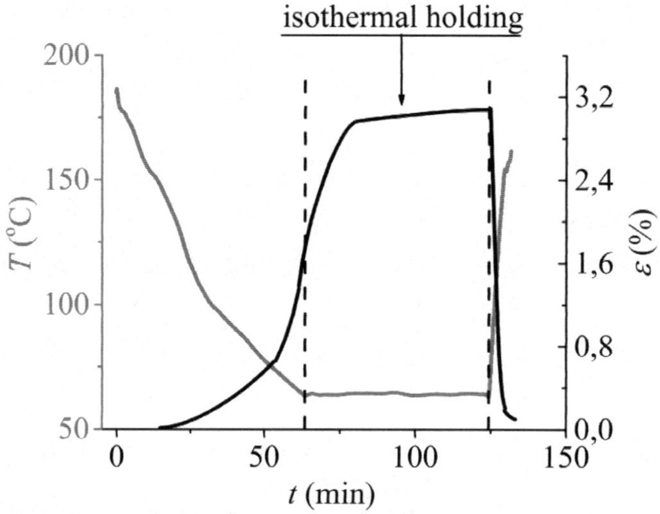

Figure 1. Temperature (red curve) and strain (black curve) variation on time during cooling, isothermal holding and heating of the $Ti_{40,7}Hf_{9,5}Ni_{41,8}Cu_8$ alloy under a stress of 400 MPa.

3. Results and discussion

Figure 2a presents the strain variation during holding at a temperature close to M_s under various constant stress. It is seen that strain increased to saturation, which value depended on stress and the maximum isothermal strain variation was found on holding under a stress of 240 MPa. Figure 2b shows the dependences of the isothermal strain variation for 60 minutes ε_{iso}^{60} on ΔT value which was equal to difference between the start temperature of the forward martensitic transformation under a stress M_s^σ (38 °C at 160 MPa, 48 °C at 240 MPa and 83 °C at 400 MPa) and holding temperature T^*. It can be seen that these dependences were non-monotonic and the maximum was observed at ΔT = -6±1 °C for all curves. The value of maximum isothermal stain depended on the stress and the largest value of 3,2 % was found on holding under 240 MPa.

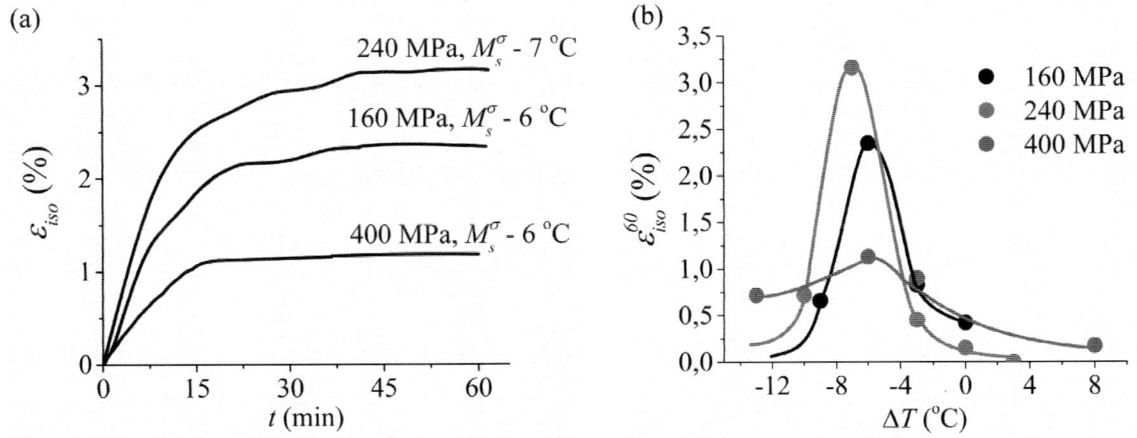

Figure 2. Strain variation during isothermal holding under various stress (a) and the dependences of isothermal strain variation for 60 min on ΔT under different stress (b).

Results found in the present work for the $Ti_{40,7}Hf_{9,5}Ni_{41,8}Cu_8$ alloy were compared with data obtained previously for the $Ti_{40,7}Hf_{9,5}Ni_{44,8}Cu_5$ alloy in [9]. In both alloys, holding under a stress at a temperature belonging to a temperature range of the forward martensitic transformation was accompanied by the isothermal strain variation that was completely recoverable on subsequent heating. The dependences of the isothermal strain on the position of the holding temperature compared to the M_s^σ temperature or on the stress were non-monotonic and the optimal regime (stress and holding duration) could be found for each alloy. The maximum (ε_{iso}^{max}) value was observed on holding at various stress in the alloys with the different Cu concentration (Figure 3). In the $Ti_{40,7}Hf_{9,5}Ni_{44,8}Cu_5$ alloy, the maximum isothermal strain was equal to 3,5 % and found on holding under 160 MPa whereas, ε_{iso}^{max} was equal to 3,2 % and observed on holding under 240 MPa in the $Ti_{40,7}Hf_{9,5}Ni_{41,8}Cu_8$ alloy.

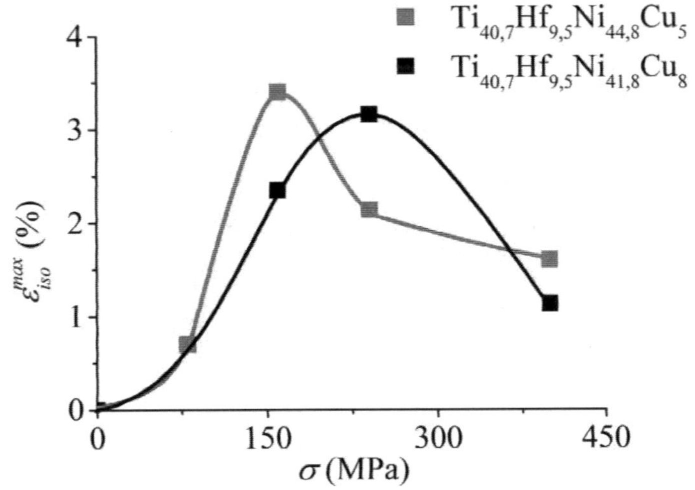

Figure 3. Dependence of the maximum isothermal strain on stress acted on holding stress.

Therefore, the observation of the strain variation on holding under a stress in the Ti-Hf-Ni-Cu alloys with various concentration of Cu and Ni atoms shows that this phenomenon is common feature of the alloys undergoing the isothermal martensitic transformations. All strain accumulated on holding completely recovers on subsequent heating hence, this strain variation is caused by the isothermal formation of martensitic under a constant stress. The variation in Ni/Cu ratio in $Ti_{40,7}Hf_{9,5}Ni_{49,8-x}Cu_x$ alloys influences the stress at which the maximum isothermal strain variation is observed. The variation in the Cu concentration slightly decreases the maximum isothermal strain from 3,5 % in the $Ti_{40,7}Hf_{9,5}Ni_{44,8}Cu_5$ alloy to 3,2 % in the $Ti_{40,7}Hf_{9,5}Ni_{41,8}Cu_8$ alloy however, to determine the influence of the alloy composition of the isothermal strain it is necessary to study the isothermal strain variation in the alloys with Cu concentration of 10 and more that will be carried out in future.

4. Conclusions

- A holding of the $Ti_{40,7}Hf_{9,5}Ni_{41,8}Cu_8$ alloy under a stress at temperatures belonging to the temperature range of the forward martensitic transformation was accompanied by the strain variation which was completely recovered on subsequent heating.
- The dependences of the isothermal strain on holding temperature or stress were non-monotonical and optimal regimes of holding can be found.
- The maximum isothermal strain was found on holding of $Ti_{40,7}Hf_{9,5}Ni_{41,8}Cu_8$ alloy under a stress of 240 MPa and this value was equal to 3,2 % that was less than in the $Ti_{40,7}Hf_{9,5}Ni_{44,8}Cu_5$ alloy.

Acknowledgment
This work was supported by Russian Scientific Foundation (grant number 18-19-00226-P)

References

[1] Otsuka K and Ren X 2005 *Progr Mater Sci* **50** 511-678
[2] Otsuka K, Ren X and Takeda T 2001 *Scr Mater* **45(2)** 145-152
[3] Akamine H, Soejima Y, Nakamura T, Sahar F, Fukuda T, Kakeshita T and Nishida M 2020 *Mater Trans* **61(1)** 37-41
[4] Ji Y, Wang D, Ding X, Otsuka K and Ren X 2015 *Phys Rev Let.* **114** 055701
[5] Kustov S, Salas D, Cesari E, Santamarta R and Van Humbeeck J 2012 *Acta Mater* **60** 2578-92
[6] Demidova E, Belyaev S, Resnina N and Shelyakov A 2020 *J Therm Anal Calorim* **139** 29665-70
[7] Resnina N, Belyaev S, Demidova E, Ivanov A and Andreev V 2018 *Mater Lett* **228** 348-350
[8] Ivanov A, Belyaev S, Resnina N and Andreev V 2019 *Sensor Actuat A - Phys* **297** 111543
[9] Demidova E, Belyaev S, Resnina N and Shelyakov A 2019 *Mater Lett* **254**, 266-8

Ab initio study of the phase stability of modulated structures in Co-doped Ni-Mn-In (Sn) Heusler alloys

K R Erager*, V V Sokolovskiy and V D Buchelnikov

Chelyabinsk State University, 454001 Chelyabinsk, Russia

*Corresponding author: eragerk@rambler.ru

Abstract

Using *ab initio* calculations, the phase stability of modulated and tetragonal martensitic structures in $Ni_{43.75}Co_{6.25}Mn_{43.75}(In, Sn)_{6.25}$ Heusler alloys with different magnetic order is investigated. The stability against the segregation is considered by a method for generating all possible decay reactions assuming the calculated ground state energies of each composition. It is shown that the highest probable stability under equilibrium conditions is demonstrated by alloys with tetragonal martensitic structure in accordance with reactions: $Ni_{35}Co_5Mn_{35}In_5 \rightarrow 25Mn + 35Ni + 5Mn_2InCo$ and $Ni_{35}Co_5Mn_{35}Sn_5 \rightarrow 5CoSn + 35Mn + 35Ni$.

Keywords: Heusler alloys, ab initio calculations, modulated martensitic structures, segregation and phase stability.

1. Introduction

Nowadays, the comprehensive research, fabrication and integration of a new class of functional materials are main goal of modern material science. Among the promising candidates, Mn-rich $Ni_{2-x}Co_xMn_{1+y}Z_{1-y}$ (Z = In, Sn, Sb) Heusler alloys attract a particular attention due to exhibiting of various multifunctional properties such as the metamagnetic shape memory effect, the giant inverse magnetocaloric effect [1, 2], the exchange bias effect [3, 4] and etc. These unique properties associated with magnetic and structural phase transformations that yields a wide spectrum of potential applications as a new generation of intelligent sensors and adapters as well as magnetic cooling devices [5]. In a case of parent $Ni_{50}Mn_{50-x}Z_x$ compounds, the martensitic phase transformation caused by temperature or magnetic field occurs usually in a certain range of compositions ($x \leq 16$ for Z = In [6] and $x \leq 15$ for Z = Sn [7]) and in a narrow temperature range. The partial substitution of Ni by Co usually leads to reduce the martensitic transition temperature and to increase both the Curie temperature and saturation magnetization of austenite. As consequence, the Zeeman energy is increased also. For example, $Ni_{45}Co_5Mn_{36.7}In_{13.3}$ compound demonstrates a field-induced transformation from the antiferromagnetic (AFM) (or paramagnetic (PM)) martensitic phase to the ferromagnetic (FM) austenitic phase at 298 K [8].

It is important to note that the research of phase stability is one of the most important issues in the study of Heusler compounds. However, not enough attention has been paid to this topic. Yuhasz *et al.* [9] reported about the heat treatment dependence on the stability of Mn-excess Heusler alloys. It has been shown that the segregation in $Ni_{50}Mn_{39}Sn_{11}$ occurs after a two-stage heat treatment at 1200 and 773 K. Recent experimental data [10-13] showed that AFM Heusler alloys $Ni_{50}Mn_{45}Z_5$ (Z = Sn, Al, Ga, In) were also unstable and decompose into two-phase FM $L2_1$-cubic $Ni_{50}Mn_{25}Z_{25}$ and AFM $Ni_{50}Mn_{50}$ with $L1_0$-tetragonal phase at annealing temperature about 650 – 800 K. Sokolovskiy *et al.* [14] carried out *ab initio* study of phase stability in a series of Heusler alloys Ni_2Mn_{1+x}(In, Sn, Ga, Al)$_{1-x}$, Mn_2Ni_{1+x} (Ga, Al) and $Ni_{2+x}Mn_{1-x}$Ga. They concluded that Ni_2Mn_{1+x}(In, Sn, Al)$_{1-x}$ compounds with Mn-excess content at low temperatures were unstable in the entire range of concentrations and

IOP Conf. Series: Materials Science and Engineering 1213 (2022) 012008 doi:10.1088/1757-899X/1213/1/012008

decompose into a two-phase system. However, $Ni_2Mn_{1+x}Ga_{1-x}$ and $Mn_2Ni_{1+x}(Ga, Al)_{1-x}$ are predicted to be stable in a narrow concentration range around stoichiometry 2-1-1, while FM $Ni_{2+x}Mn_{1-x}Ga$ alloys are stable in the concentration range $0 < x < 0.6$. The structural characteristics of martensite in Ni-Mn-In alloy have been extensively studied in Refs. [15-17]. Depending on the composition, the martensitic phase could be a five-layer modulated (5M), six-layer modulated (6M), seven-layer modulated (7M) and non-modulated (NM) martensite with $L1_0$-tetragonal or orthorhombic structure. Eventually, a modulated martensite transforms into a NM one at low temperatures.

The functional properties of FM shape memory materials are usually achieved through the formation of a modulated martensitic structure. Therefore, it is necessary that 5M, 6M or 7M martensites will be stabilized over a wide temperature range. In this paper, we focus on the segregation effects of modulated martensitic structures in Ni-Co-Mn-(In, Sn) Heusler alloys in the framework of *ab initio* calculations.

2. Calculation details

Ab initio calculations of the ground state properties of $Ni_{43.75}Co_{6.25}Mn_{43.75}Sn_{6.25}$ and $Ni_{43.75}Co_{6.25}Mn_{43.75}In_{6.25}$ alloys were performed using the augmented projection wave (PAW) method implemented in the VASP software package [18, 19]. To describe the exchange-correlation energy, a generalized gradient approximation applying Perdew-Burke-Ernzerhof as correlation functional was considered [20]. The Brillouin zone was integrated by the Monkhorst-Pack method using a $12 \times 12 \times 12$ k-point mesh. The plane wave cutoff energy was 460 eV, and the energy convergence parameter was 10^{-8} eV/atom. Geometric optimization of the crystal structures of the austenite and martensite phases was performed within the framework of electronic and ionic relaxation. The crystal structures of the cubic austenite and NM martensite were described by the 16-atom cubic and tetragonal supercells (space groups No. 225 and 139, correspondingly). The Mn excess and Co atoms were located at In (Sn) and Ni sublattices. Therefore, nominal compositions $Ni_{43.75}Co_{6.25}Mn_{43.75}(In, Sn)_{6.25}$ can be presented in the supercell approach as $Ni_7Co_1Mn_7(In, Sn)_1$. The two magnetic orders were considered as follows: FM (all atoms have parallel aligned magnetic moments) and ferrimagnetic (FIM, magnetic moments of Mn excess atoms are reversed to other atoms) as shown in Fig. 1.

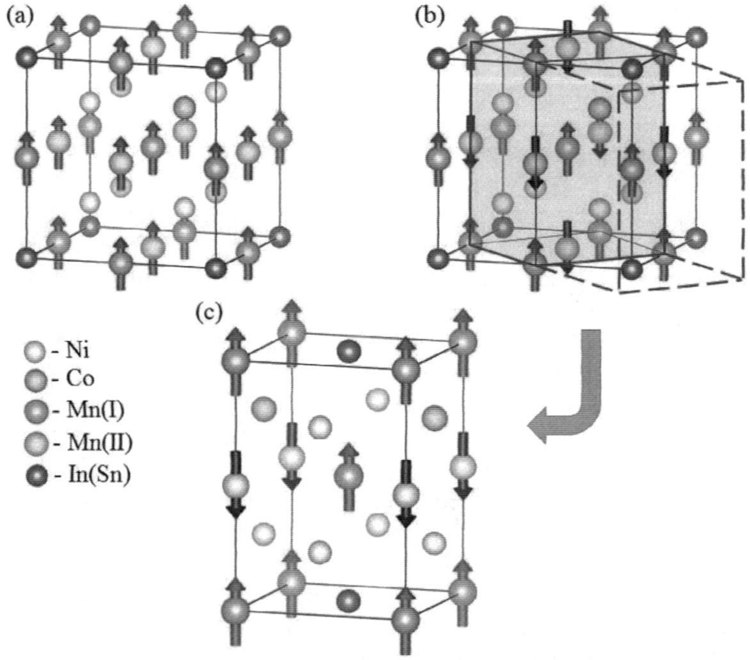

Figure 1. Cubic and tetragonal structures of austenitic and martensitic phases of $Ni_{43.75}Co_{6.25}Mn_{43.75}(In, Sn)_{6.25}$ alloys with FM and FIM order, respectively. Lattice constants a_t and a_c for tetragonal and cubic structure are related by way of $a_t = a_c/\sqrt{2}$.

IOP Conf. Series: Materials Science and Engineering 1213 (2022) 012008 doi:10.1088/1757-899X/1213/1/012008

The modulated 10M martensitic structure was described by 80-atom supercell ($Ni_{35}Co_5Mn_{35}(In, Sn)_5$) with FM and FIM order, which consists 10 tetragonal unit cells arranged in the [110] direction (see Fig. 2). An initial modulation amplitude of 0.2 Å was used for all atomic planes. Geometric optimization of 10M structure was performed using ionic relaxation method. The integration of the Brillouin zone was carried out on a Γ-centered k-point mesh (12 × 2 × 10).

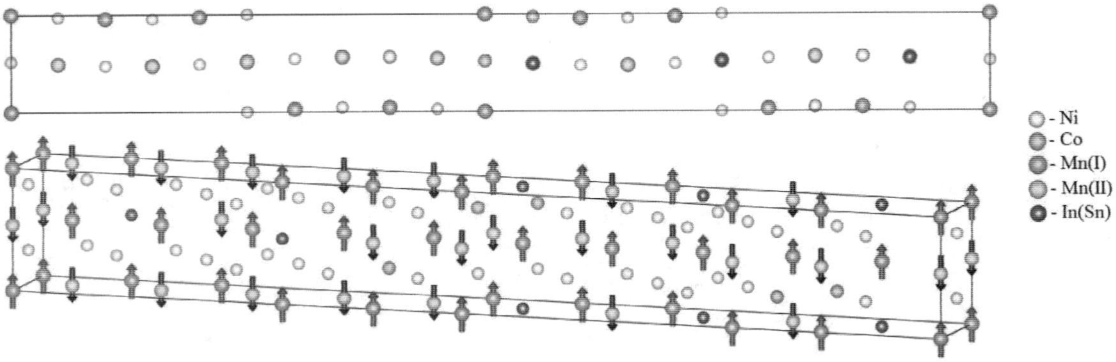

Figure 2. Crystal structure of 10M martensitic structure of $Ni_{43.75}Co_{6.25}Mn_{43.75}(In, Sn)_{6.25}$ with FIM order.

3. Results and discussions

Let us firstly discuss the results of the ground state energy calculations in the cubic austenitic and tetragonal martensite phases. Fig. 3 illustrates the total energy difference ΔE as a function of tetragonal distortion (*c/a*) in $Ni_{43.75}Co_{6.25}Mn_{43.75}(In, Sn)_{6.25}$ compounds with FM and FIM ordering of magnetic moments.

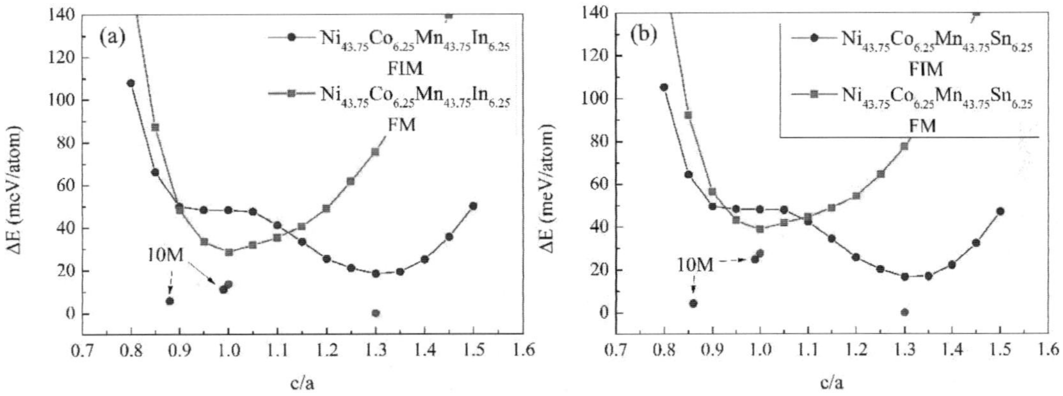

Figure 3. The energy difference *(ΔE)* as a function of tetragonal distortion (*c/a*) in $Ni_{43.75}Co_{6.25}Mn_{43.75}(In, Sn)_{6.25}$ ordered ferro- and ferrimagnetically. The dots denote the results for 10M martensitic structure. Lines with symbols are the results of electronic relaxation, dots are the results of ionic relaxation.

It is seen that for both compounds, the FM configuration is preferable in the austenitic phase with a high enough total magnetic moment (~ 7.3 – 7.4 μ_B/f.u.), while the martensitic phase with tetragonal structure (*c/a* = 1.3) reveals the antiparallel spin alignment of Mn excess atoms. As consequence, the total magnetic moment of martensitic phase is calculated to be about 0.98 μ_B/f.u. The similar observation can be found for the modulated structure of each compound. The values of the total magnetic moment of modulated and NM structures are close to each other, the difference is about ~0.037 and 0.004 μ_B/f.u. for $Ni_{43.75}Co_{6.25}Mn_{43.75}In_{6.25}$ and ~0.063 and 0.034 μ_B/f.u. for

IOP Conf. Series: Materials Science and Engineering 1213 (2022) 012008 doi:10.1088/1757-899X/1213/1/012008

$Ni_{43.75}Co_{6.25}Mn_{43.75}Sn_{6.25}$ with FM- and FIM ordering of magnetic moments, respectively. These findings are confirmed by Grünebohm et al. [21], they suggested theoretically that the presence of modulation does not affect the value of the total magnetic moment for compounds under consideration. Besides, our calculations have shown that the 10M structure turns to be close in energy with the ground state NM tetragonal martensite for both $Ni_{43.75}Co_{6.25}Mn_{43.75}(In, Sn)_{6.25}$ alloys. The analysis of optimized structural parameters of modulated structure reveals that 10M structure with initial $a=b\neq c$ parameters is converged to the monoclinic structure during a full geometry optimization calculations. The calculated partial and total magnetic moments as well as optimized lattice constants for $Ni_{43.75}Co_{6.25}Mn_{43.75}(In, Sn)_{6.25}$ with different magnetic order are summarized in Tables 1 and 2.

Table 1. The equilibrium values of partial and total magnetic moments as well as the energy difference between the considered structures in $Ni_{43.75}Co_{6.25}Mn_{43.75}(In, Sn)_{6.25}$ with respect to the energetically preferable NM tetragonal martensite with FIM order.

	Magnetic moment (μ_B/f.u.)							ΔE
	c/a	Ni	Co	Mn (I)	Mn (II)	In (Sn)	Total	(meV/atom)
$Ni_{43.75}Co_{6.25}Mn_{43.75}In_{6.25}$								
FM	1	0.705	1.472	3.319	3.335	-0.072	7.404	13.657
	0.99 (10M)	0.703	1.453	3.308	3.311	-0.073	7.367	11.180
FIM	1.33	0.102	0.358	3.002	-3.042	-0.046	0.977	0
	0.88 (10M)	0.097	0.269	3.003	-2.997	-0.050	0.981	5.822
$Ni_{43.75}Co_{6.25}Mn_{43.75}Sn_{6.25}$								
FM	1	0.667	1.466	3.320	3.364	-0.065	7.361	27.478
	0.99 (10M)	0.659	1.427	3.306	3.335	-0.074	7.298	24.753
FIM	1.33	0.108	0.481	2.976	-3.026	-0.064	0.999	0
	0.86 (10M)	0.094	0.402	2.958	-2.986	-0.070	0.965	4.310

Table 2. Optimized lattice constants and tetragonal ratio (*c/a*) for the considered structures in $Ni_{43.75}Co_{6.25}Mn_{43.75}(In, Sn)_{6.25}$ with FM and FIM order.

Compound		Lattice constants				Angle
$Ni_{43.75}Co_{6.25}Mn_{43.75}In_{6.25}$		*c/a*	*a* (Å)	*b* (Å)	*c* (Å)	$\gamma°$
austenite	FM	1	5.886	5.886	5.886	90°
NM martensite	FIM	1.33	5.272	5.271	7.037	90°
10M martensite	FM	0.99	4.161	4.170	5.841	89.31°
	FIM	0.88	4.263	4.323	5.331	89.95°
$Ni_{43.75}Co_{6.25}Mn_{43.75}Sn_{6.25}$						
austenite	FM	1	5.887	5.887	5.887	90°
NM martensite	FIM	1.33	5.240	5.240	7.117	90°
10M martensite	FM	0.99	4.160	4.168	5.852	90.38°
	FIM	0.86	4.330	4.275	5.286	89.06°

Next, we turn to a discussion of the assessment of the phase stability of the compounds under study. For each composition, all possible combinations of decomposition reactions into stable components with the corresponding stoichiometric coefficients were generated. For this, we compared the total energies of austenitic and martensitic $Ni_{43.75}Co_{6.25}Mn_{43.75}(In, Sn)_{6.25}$ and the total energies of all possible stable compositions such as Co, Co_3Ni, In, In_3Co, In_7Ni_3, InNi, Mn, Mn_3Co, MnCo, $MnNi_3$, Ni, Mn_2InCo, Ni_2MnIn, Mn_8Ni_8 for Ni-Co-Mn-In, and Co, Co_3Ni, CoSn, $CoSn_3$, Mn, Mn_3Co, MnCo, Co_2MnSn, $MnNi_3$, Ni, Ni_3Sn, Ni_3Sn_2, Ni_3Sn_4, Mn_8Ni_8 for Ni-Co-Mn-Sn. The list of stable binary compounds was taken from the "Materials project" database. As a result, we calculated the decomposition energy E_{dec} by the following way

$$E_{dec} = E_0^{\text{Ni-Co-Mn-(In,Sn)}} - \sum_i^N E_i^{stable}, \quad (1)$$

where E_0 is the total energy of compounds under study, E_i is the total energy of each component in its stable crystal structure, N is a number of decay components. The negative value of E_{dec} indicates that a given compound will be stable against the segregation into a mixture of binary compounds and pure elements and *vice versa*. It means that a given compound can be synthetized under equilibrium conditions. It should be noted that the heat treatment of nonstoichiometric compound can lead to the disappearance of martensitic transformation and to the appearance of segregation into systems consisting of two or more phases as follows from experimental findings [9-13].

In Figs. 4 and 5 we present the results of decomposition energy calculations for $Ni_{43.75}Co_{6.25}Mn_{43.75}(In, Sn)_{6.25}$ in austenitic and martensitic phase ordered ferro- and ferrimagnetically.

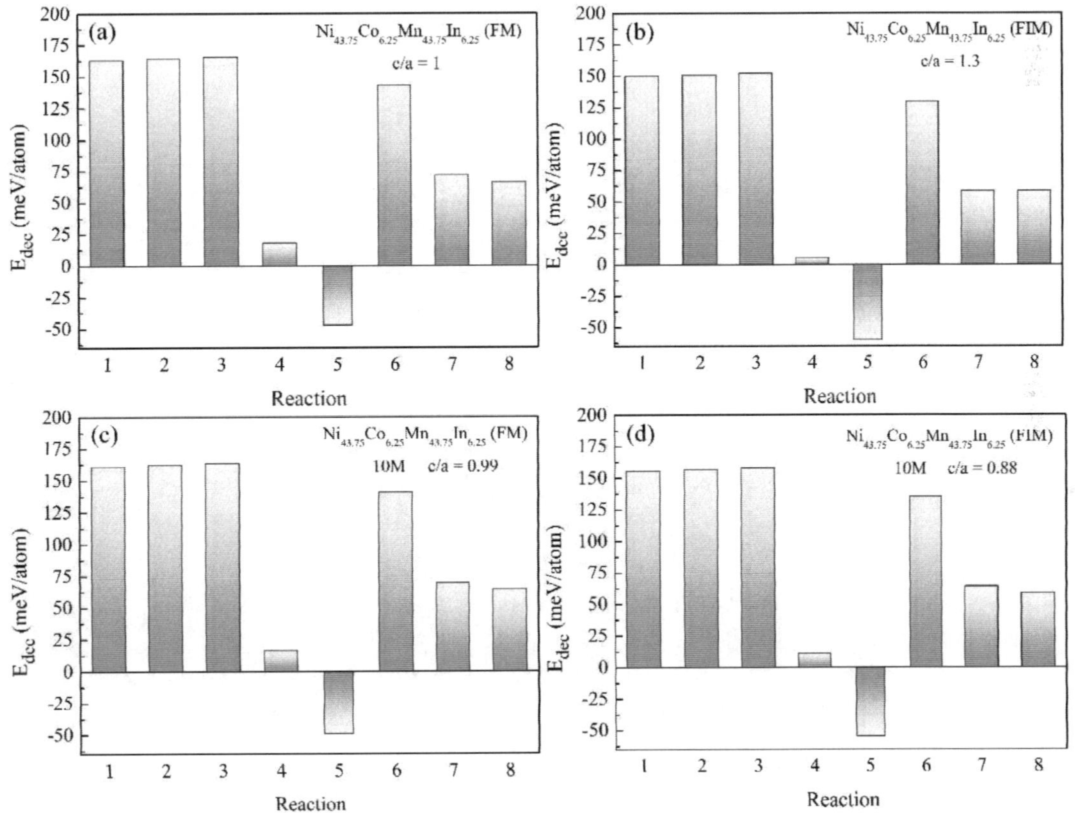

Figure 4. Decomposition energy for $Ni_{43.75}Co_{6.25}Mn_{43.75}In_{6.25}$ with (a) FM austenite, (b) FIM NM martensite, (c, d) FM and FIM 10M martensite. The list of generated decomposition reactions is as follows: (#1) $8Ni_{35}Co_5Mn_{35}In_5 \rightarrow 40Co + 40In + 35Mn_8Ni_8$; (#2) $24Ni_{35}Co_5Mn_{35}In_5 \rightarrow 80Co + 40In_3Co_1 + 105Mn_8Ni_8$; (#3) $4Ni_{35}Co_5Mn_{35}In_5 \rightarrow 20InNi + 20MnCo + 15Mn_8Ni_8$; (#4) $3Ni_{35}Co_5Mn_{35}In_5 \rightarrow 40Mn + 35MnNi_3 + 15Mn_2InCo$; (#5) $Ni_{35}Co_5Mn_{35}In_5 \rightarrow 25Mn + 35Ni + 5Mn_2InCo$; (#6) $8Ni_{35}Co_5Mn_{35}In_5 \rightarrow 40MnCo + 40Ni_2MnIn + 25Mn_8Ni_8$; (#7) $2Ni_{35}Co_5Mn_{35}In_5 \rightarrow 10Mn1Ni_3 + 10Mn_2InCo + 5Mn_8Ni_8$; (#8) $8Ni_{35}Co_5Mn_{35}In_5 \rightarrow 80Ni + 40Mn_2InCo + 25Mn_8Ni_8$.

It is obvious that the most of considered reactions give the positive value of the decomposition energy, which indicates the possible decomposition processes in $Ni_{43.75}Co_{6.25}Mn_{43.75}(In, Sn)_{6.25}$. Nevertheless, a high probable stability for all considered structures of $Ni_{43.75}Co_{6.25}Mn_{43.75}In_{6.25}$ is found for the reaction #5. In this case, the largest negative value of E_{dec} is observed for NM martensitic structure. For $Ni_{43.75}Co_{6.25}Mn_{43.75}Sn_{6.25}$, the preferable reaction giving the stability of both 10M and NM structures against decomposition is the reaction #4. We suggest that $Ni_{43.75}Co_{6.25}Mn_{43.75}In_{6.25}$ compound can be metastable in both austenite and martensite phase. In regard to $Ni_{43.75}Co_{6.25}Mn_{43.75}Sn_{6.25}$, it reveals the phase metastability in the martensitic structure while the austenitic phase is unstable and decomposed into a mixture of different stable binary compounds. As can be seen from Figs. 4 (b, d) and 5(b, d), the decomposition energy difference between 10M and NM structures is found to be small and equal to ~5.8 and ~4 meV/atom for $Ni_{43.75}Co_{6.25}Mn_{43.75}In_{6.25}$ and $Ni_{43.75}Co_{6.25}Mn_{43.75}Sn_{6.25}$, respectively. This finding suggests that 10M and NM structures can coexist together in martensitic phase.

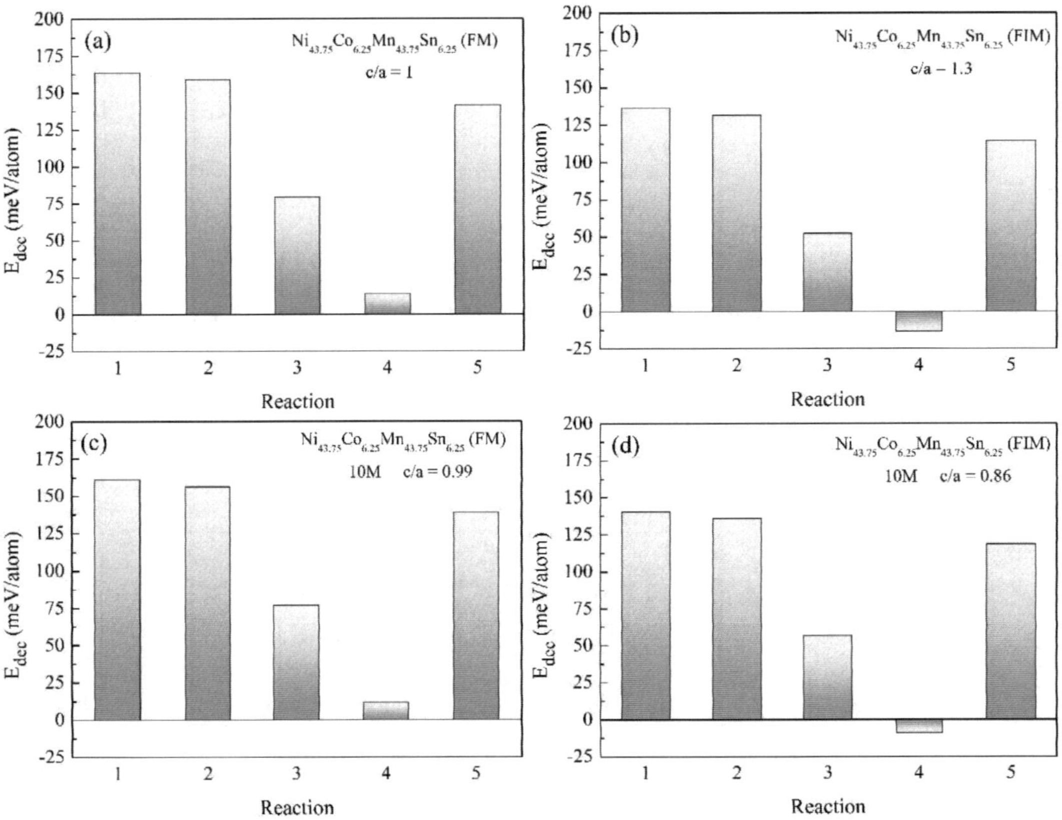

Figure 5. Decomposition energy for $Ni_{43.75}Co_{6.25}Mn_{43.75}Sn_{6.25}$ with (a) FM austenite, (b) FIM NM martensite, (c, d) FM and FIM 10M martensite. The list of generated decomposition reactions is as follows: (#1) $24Ni_{35}Co_5Mn_{35}Sn_5 \rightarrow 80Co + 40CoSn_3 + 105Mn_8Ni_8$; (#2) $8Ni_{35}Co_5Mn_{35}Sn_5 \rightarrow 40Co + 40Sn + 35Mn_8Ni_8$; (#3) $3Ni_{35}Co_5Mn_{35}Sn_5 \rightarrow 15CoSn + 70Mn + 35MnNi_3$ (#4) $Ni_{35}Co_5Mn_{35}Sn_5 \rightarrow 5CoSn + 35Mn + 35Ni$; (#5) $2Ni_{35}Co_5Mn_{35}Sn_5 \rightarrow 10Mn_3Co + 10Ni_3Sn + 5Mn_8Ni_8$.

4. Conclusions

This work presents the results of *ab initio* study of the phase stability in the $Ni_{43.75}Co_{6.25}Mn_{43.75}In_{6.25}$ and $Ni_{43.75}Co_{6.25}Mn_{43.75}Sn_{6.25}$ Heusler alloys with different types of magnetic ordering. For both compounds, it is shown that monoclinic modulated 10M structure with ferrimagnetic ordering of magnetic moments is close in energy to the NM martensitic phase. In turn, the period and amplitude of modulation structure affect no the total magnetic moment of martensitic phase; the difference between the magnetic moments for NM and 10M martensite takes ~ 0.004 and ~ 0.063 μ_B/f.u. in a case of $Ni_{43.75}Co_{6.25}Mn_{43.75}In_{6.25}$ and $Ni_{43.75}Co_{6.25}Mn_{43.75}Sn_{6.25}$, respectively. The practically equal total energies of NM and 10M martensitic structures of $Ni_{43.75}Co_{6.25}Mn_{43.75}In_{6.25}$ alloy indicate that the NM and 10M structures in this alloy can coexist; however, compositions with the structure of NM martensite demonstrate a greater probable stability.

Acknowledgment

Calculations of modulated structures was carried out with the support of the Russian Science Foundation project No. 17-72-20022. Calculations of the structure phase stability was supported by the Ministry of Science and Higher Education of the Russian Federation within the framework of the Russian State Assignment under contract No. 075-00992-21-00.

References

[1] Liu J, Gottschall T, Skokov K P, Moore J D and Gutfleisch O 2012 *Nat. Mater.* 11 620.
[2] Dubenko I, Khan M, Pathak A K, Gautam B R, Stadler S and Ali N 2009 J. *Magn. Magn. Mater.* 321 754.
[3] Wang B M, Liu Y, Wang L, Huang S L, Zhao Y, Yang Y and Zhang H 2008 *J. Appl. Phys.* 104 043916.
[4] Wang B M, Liu Y, Ren P, Xia B, Ruan K B, Yi J B, Ding J, Li X G and Wang L 2011 *Phys. Rev. L.* 106 077203.
[5] Umetsu R Y, Xu X and Kainuma R 2016 *Scr. Mater.* 116 1.
[6] Krenke T, Acet M, Wassermann E, Moya X and Mañosa L 2006 *A. Planes, Phys. Rev. B* 73 174413.
[7] Krenke T, Acet M, Wassermann E, Moya X and Mañosa L 2005 *A. Planes, Phys. Rev. B* 72 014412.
[8] Kainuma R, Imano Y, Ito W, Sutou Y, Morito H, Okamoto S, Kitakami O, Oikawa K, Fujita A, Kanomata T and Ishida K 2006 *Nature.* B 439 957.
[9] Yuhasz W M, Schlagel D L, Xing Q, McCallum R W., and Lograsso T A 2010 *J. Alloys Compd.* 492 681.
[10] Krenke T, Çakir A, Scheibel F, Acet M, and Farle M 2016 *J. Appl. Phys.* 120 243904.
[11] Çakir A, Acet M, and Farle M 2016 *Sci. Rep.* 6 28931.
[12] Çakir A and Acet M 2017 *AIP Adv.* 7 056424.
[13] Çakir A, Acet M, Wiedwald U, Krenke T, and Farle M 2017 *Acta Mater.* 127 117.
[14] Sokolovskiy V V, Gruner M E, Entel P, Acet M, Cakir A, Baigutlin D R and Buchelnikov V D 2019 *Phys. Rev. Mat.* V 3. 084413.
[15] Ito W, Imano Y, Kainuma R, Sutou Y, Oikawa K and Ishida K 2007 *Mater. Trans. A.* 38 759-766.
[16] Abematsu K, Umetsu R Y, Kainuma R, Kanomata T, Watanabe K and Koyama1 K 2014 *Mater. Trans.* 55 477-481.
[17] Krenke T, Duman E, Acet M, Wassermann E, Moya X, Manosa L, Planes A, Suard E and Ouladdiaf B 2007 *Phys. Rev. B.* 75 104414.
[18] Kresse G and Furthmüller J 1996 *Phys. Rev. B* 54 11169.
[19] Kresse G and Joubert D 1999 *Phys. Rev. B* 59 1758.
[20] Perdew J P, Burke K, and Ernzerhof M 1996 *Phys. Rev.* Lett. **77** 3865.

[21] Grünebohm A, Herper H C and Entel P 2016 J. *Phys. D: Appl. Phys.* 49 395001.

IOP Conf. Series: Materials Science and Engineering 1213 (2022) 012009 doi:10.1088/1757-899X/1213/1/012009

Investigation of crystal structure formation by electropulse treatment of amorphous $Ti_{50}Ni_{25}Cu_{25}$ alloy ribbons

N Sitnikov[1,2*], A Shelyakov[2] and I Zaletova[1]

[1] The State Scientific Centre "Keldysh Research Center", Onezhskaya St. 8, 125438 Moscow, Russia
[2]National Research Nuclear University MEPhI (Moscow Engineering Physics Institute), 31 Kashirskoe shosse, Moscow 115409, Russia

*Corresponding author e-mail: sitnikov_nikolay@mail.ru

Abstract. The study of the effect of electropulse treatment with a variable duration on the crystallization processes and the structure of a amorphous TiNiCu alloy with 25 at.% Cu in comparison with isothermal annealing and heating at a constant speed was carried out. The alloy was fabricated by rapid-quenching from the liquid state (melt spinning technique) at the cooling rate of the melt of about 10^6 °C/s in the form of a ribbon with a thickness of 28 μm with a surface crystal layer with a thickness of about 2-3 μm. To remove the crystal layer, the method of double-sided electrochemical polishing was used. The studies were carried out by methods of differential scanning calorimetry, metallography and scanning electron microscopy. It was established that the formation of the crystalline phase in the electropulse treatment of the amorphous ribbon occurs from the surface to the inner part due to the predominant formation and growth of columnar crystals with subsequent nucleation and growth of crystals in the rest of the ribbon.

1. Introduction

Copper-doped TiNi alloys attract attention as materials with the shape memory effect (SME) due to the narrow temperature hysteresis of the SME manifestation and the relatively large return deformation [1-4]. Recently, the prospects of using alloys of the quasi–binary TiNi-TiCu system obtained by rapid-quenching from a melt in the form of thin ribbons have been shown to create microactuators, in particular, micropincers, which are a micromechanical tool that provides capture and retention for spatial manipulation and preparation of micro- and nanoobjects [5-7]. Previously, it was found that TiNiCu alloys crystallized from the amorphous state have the best characteristics, and amorphization is achieved in alloys with a high copper content (more than 20 at. %) at a melt cooling rate of about 10^6 °C/s [3]. To use amorphous alloys as functional materials with SME, it is necessary to crystallize the alloys using heat treatment. Recently, many studies have shown that the phase composition, morphology and average grain size of alloys crystallized from the amorphous state strongly depends on the crystallization conditions. Annealing with an ultra-high heating rate (for example, electric pulse treatment) has shown significant advantages over traditional isothermal heat treatment in many aspects [8]. High-density electrical pulses cause structural changes in metals and alloys, such as dynamic recrystallization, dislocation movements, grain grinding and the formation of oriented microstructures, which can lead to a significant improvement in the physical and mechanical

IOP Conf. Series: Materials Science and Engineering 1213 (2022) 012009 doi:10.1088/1757-899X/1213/1/012009

properties of materials. In particular, it is shown that electric pulse treatment improves the characteristics of SME and superelasticity in TiNi-based alloys [9-12]. It was recently established [13] that high-speed electric pulse annealing with a duration of 10 ms provides a bright manifestation of SME in alloys of the TiNi-TiCu system with a copper content of up to 38 at. %, which after isothermal annealing become brittle and are not able to exhibit SME. The aim of this work is to study the effect of electric pulse treatment with a variable duration on the crystallization processes and the structure of a rapidly-quenched amorphous TiNiCu alloy with 25 at. % Cu in comparison with isothermal annealing and heating at a constant speed.

2. Materials and methods

In this work, we studied thin ribbons made of $Ti_{50}Ni_{25}Cu_{25}$ alloy (at. %) obtained by rapid-quenching from a liquid state on a fast-rotating copper disk (melt spinning technique) [5, 6]. During the quenching process, the melt cooling rate was about 10^6 °C/s, which ensured the production of a ribbon with a thickness of 25-40 μm in a predominantly amorphous state with a thin (several microns) surface crystal layer on the non-contact side of the ribbon relative to the quenching disk. The selected melt cooling rate allows us to obtain a ribbon with sufficiently uniform parameters in width and thickness, which is a necessary condition for uniform heating of the ribbon by passing a pulsed electric current during further crystallization. A large degree of amorphization of the alloy is achieved due to an increase in the cooling rate of the melt with an increase in the rotation speed of the quenching disk, but this leads to a decrease in the quality of the ribbon (jagged edges, uneven thickness, through holes). The obtained samples of the rapid-quenched $Ti_{50}Ni_{25}Cu_{25}$ alloy were extended ribbons (up to 20 m) with an average thickness of about 28 μm and a width near to 1.5 mm.

To remove the crystal layer and further smooth the surface relief, the method of double-sided electrochemical polishing (ECP) was used with immersion of the ribbon samples in the electrolyte of the "PLS-5" brand produced by "STC Technocom AS". ECP was carried out under normal conditions at a voltage of 5 V and a current density of 30–50 mA/cm^2. At the specified ECP parameters, the sample thinning rate was approximately 0.5 μm/min. To achieve a completely amorphous state, samples of ribbons with a length of 70 mm were subjected to ECP for 12 minutes and 20 minutes. As a result of this treatment, along with controlled thinning, the roughness of the ribbon surfaces was further reduced due to the predominant etching of the protruding areas.

The crystallization of the amorphous state of rapidly-quenched amorphous ribbons made of $Ti_{50}Ni_{25}Cu_{25}$ alloy was carried out by the following methods: 1) isothermal treatment (ITT) in a furnace at a temperature above the crystallization temperature; 2) heat treatment at a constant heating rate to the crystallization temperature (speed heat treatment – SHT); 3) dynamic heat treatment with crystallization by electropulse treatment (EPT).

ITT in a muffle furnace was carried out at temperatures of 500 °C and 460 °C with exposure for 300 seconds and subsequent quenching in water.

SHT with crystallization was carried out in a differential scanning calorimeter "STA 449F1 Jupiter" at a heating rate of 10 °C/min in the temperature range from 30 °C to 480 °C in a dynamic argon atmosphere.

Dynamic heat treatment with crystallization of the amorphous state was carried out on a special laboratory installation for thin ribbons by passing a single pulse of electric current through the sample, as a result of which the sample is heated and crystallized in time intervals of several milliseconds or less [13-15]. With the correct selection of the mode, it is possible to achieve complete crystallization of the amorphous state in thin ribbons in a short time without thermal exposure.

To calculate the parameters of the electric current pulse required for heating the samples due to the release of Joule heat to the required temperature (in particular, from room temperature to the crystallization temperature), we used the previously obtained ratio linking the electric current density and the duration of exposure to the pulse [15].

To study the samples of ribbons obtained in the work, their transverse slots were made on the equipment for metallographic studies of the company "Buehler" (precision cutting machine "Isomet

IOP Conf. Series: Materials Science and Engineering 1213 (2022) 012009 doi:10.1088/1757-899X/1213/1/012009

1000", automatic hydraulic press for hot pressing "Simplimet 1000", grinding and polishing machine "EcoMet 250+AutoMet 250"). For additional manifestation of the structure, etching of the polished surface with a solution of HF (5 %) + HNO_3 (25 %) + H_2O (70 %) for 10-30 seconds was used, followed by wiping with a 3 % H_2O_2 solution to remove plaque from the etching products. The microstructure of the samples was studied using an inverted metallographic reflected light microscope "Carl Zeiss Axiovert 40 MAT" and a scanning electron microscope (SEM) "FEI Quanta 600 FEG".

3. Results and discussion

The study of the cross-section of the manufactured rapidly-quenched ribbon made of $Ti_{50}Ni_{25}Cu_{25}$ alloy showed that on the non-contact surface of the ribbon (with respect to the quenching disk) there is a continuous layer of the crystalline phase with a thickness of about 2 μm with a total thickness of the ribbon of about 28 μm (Fig. 1a). After ECP, no surface crystal layer was detected, while the thickness of the thinned ribbon was about 20 μm (Fig. 1b), the length of the thinned ribbons was about 40 mm. Further, the obtained refined amorphous ribbons were subjected to heat treatment by ITT, SHT and EPT.

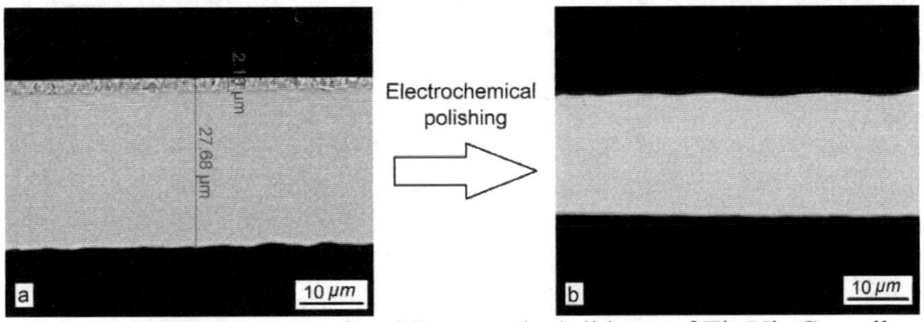

Figure 1. Cross-sectional SEM images of rapidly quenched ribbons of $Ti_{50}Ni_{25}Cu_{25}$ alloy, in the initial state (a) and after electrochemical polishing (b).

From the obtained DSC curve (Fig. 2) of the crystallization process of an amorphous ribbon made of $Ti_{50}Ni_{25}Cu_{25}$ alloy after ECP at a heating rate of 10 °C/min, it can be seen that at a temperature of about 440 °C, the amorphous state begins to decompose, and in the temperature range of 450÷460 °C, the main part of the phase transformation occurs with the formation of a crystalline state. The temperature of the peak heat release during crystallization is 454.5 °C, with a total enthalpy of the process of 31 J/g.

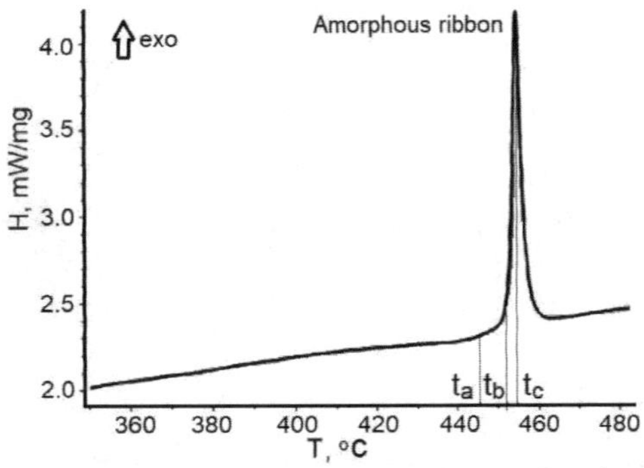

Figure 2. Heating DSC curve of amorphous $Ti_{50}Ni_{25}Cu_{25}$ alloy.

To study the crystallization process of the amorphous state at a constant heating rate, annealing was carried out in DSC to temperatures t_a=445 °C (the temperature before the start of crystallization), t_b=452 °C (the temperature after the start of crystallization) and t_c=460 °C (the temperature of the end

IOP Conf. Series: Materials Science and Engineering 1213 (2022) 012009 doi:10.1088/1757-899X/1213/1/012009

of the crystallization process). The samples were heated to the appropriate temperature, and then the heating was interrupted and the chamber was cooled.

Additionally, the crystallization of amorphous ribbons was carried out by the ITT method at a temperature of t_c=460 °C, as well as at 500 °C – the standard processing temperature for the alloy under study [13].

SEM studies of cross-sections of heat-treated ribbons (Fig. 3, 4) showed that the treatment of the amorphous state with the selected modes lead to the formation of both a fully crystallized and partially crystallized microstructure in the studied samples.

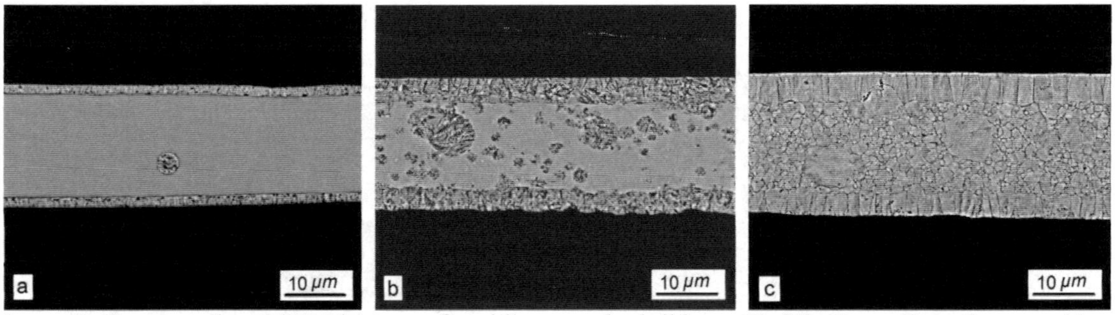

Figure 3. Cross-sectional SEM images of rapidly quenched ribbons $Ti_{50}Ni_{25}Cu_{25}$ alloy after speed heat treatment (SHT) in DSC up to temperature: t_a=445 °C (a); t_b=452 °C (b); t_c=460 °C (c).

In the structure of the cross-section of the ribbon processed by the SHT method to a temperature of t_a=445 °C, a continuous crystalline phase (a surface crystal layer) is observed near the surface, and the inner part of the ribbon has retained a predominantly amorphous state in which rare single crystals are present (Fig. 3a). When processed to a temperature of t_b=452 °C, changes occur in the structure of the ribbon, obviously associated with the development of the crystallization process – the thickness of the surface crystal layers increases, and the number of single and combined crystals and their size increases in the amorphous part (Fig. 3b). When processed to a temperature of t_c=460 °C, the sample is characterized by a completely crystallized state, and the structure of the ribbon in the cross section has an inhomogeneous distribution of crystals (Fig. 3c): columnar crystals with a characteristic size of 5-6 μm are observed along the edges of the ribbon, and micron-sized crystals are present in the inner part, among which there are single larger crystals with sizes more than 10 μm. An increase in the processing temperature to 480 °C does not significantly change the crystallized structure of the samples. Thus, during heat treatment with a constant heating rate, the formation of the crystalline phase first begins from the surface of the ribbon, followed by its growth into the inner part (in the direction of heat removal into the ribbon), and after volumetric heating of the ribbon, the crystallization of the inner part of the ribbon occurs due to the formation, growth and unification of crystallites.

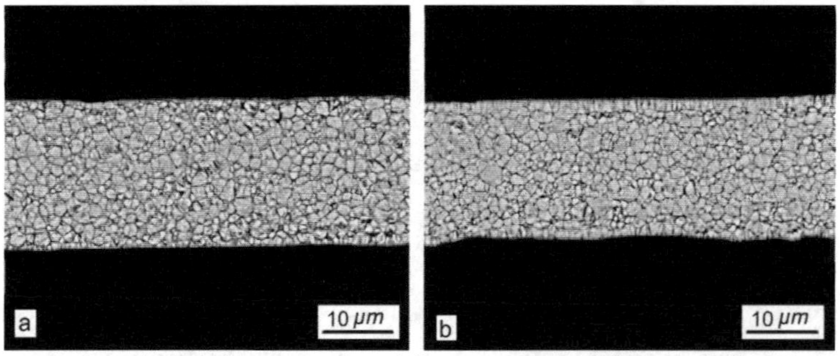

Figure 4. Cross-sectional SEM images of rapidly quenched ribbons of $Ti_{50}Ni_{25}Cu_{25}$ alloy after isothermal treatment (ITT) at temperatures 500 °C (a) and t_c=460 °C (b).

A different pattern of the structure is observed after the crystallization of amorphous ribbons by the ITT method in a muffle furnace according to the standard method at 500 °C with an exposure of 300 seconds and subsequent quenching in water. The cross-sectional structure of the heat-treated ribbon, shown in Figure 4a, is characterized by a homogeneous (uniform) structure with a characteristic crystal size of about 2-3 µm, the surface crystal layer is not detected. With such heat treatment, the ribbon is rapidly heated, which contributes to the nucleation of crystals in the entire volume of the amorphous ribbon and their uniform growth [13]

After ITT at a temperature t_c=460 °C, the cross-sectional structure of the ribbon differs from the structure obtained at 500 °C by the presence of crystalline layers with a thickness of about 2 µm at the ribbon surfaces (Fig. 4b). This is probably due to a lower temperature gradient in the ribbon during heat treatment and, accordingly, with a lower rate of heat propagation from the surface into the ribbon. It should be noted that the observed surface crystals are similar to those formed during processing by the SHT method to a temperature of t_a=460 °C, but have smaller sizes.

EPT (dynamic crystallization) radically changes the formed crystal structure in the treated $Ti_{50}Ni_{25}Cu_{25}$ alloy ribbons (Fig. 5) [13, 14]. At the same time, the processing duration (the duration of the electric current pulse) has a significant impact. With EPT for 5 seconds or more, the cross-sectional structure of the ribbon is similar to the structure formed by the ITT method at 500 °C (Fig. 5a). A decrease in the duration of the EPT leads to the formation of a structure with an inhomogeneous distribution of crystals over the volume of the ribbon: columnar crystals are present near the ribbon surfaces, and single or grouped large crystals are observed in the volume of the ribbon (Fig. 5b-d), which somewhat resembles the structure when processing by the SHT method up to 460 °C (Fig. 3c). However, there are also significant differences due to different heating and crystallization mechanisms in these two cases. This is especially evident when the duration of EPT is reduced, which leads to an increase in the proportion of surface columnar crystals on both sides of the ribbon, and columnar crystals grow inside the ribbon to different depths, while an uneven interface with crystals from the inner part is formed (Fig. 5c, d). With an EPT with a duration of 10 ms or less, the proportion of columnar crystals prevails in the structure of the ribbon, and it is possible to form local regions in which columnar crystals from opposite surfaces touch each other and form a smooth border [14].

It is noteworthy that some samples after EPT were in a state of pronounced inhomogeneous crystallization in the longitudinal direction of the ribbon. In particular, such samples were characterized by different elasticity along the ribbon and the color of oxides on the surface of the ribbon. Obviously, such a state of the ribbons after crystallization by a current pulse is associated with inhomogeneity of the cross-section and defects in the structure of the ribbon.

To study the observed phenomenon, a sample with a length of 16 mm with an EPT exposure duration of 10 ms was selected. The elastic properties of the sample change dramatically along its length from one edge to the other: one end of the ribbon has retained the appearance and elasticity of the initial amorphous state, and the other end of the ribbon has become plastic and has been coated with an oxide characteristic of the state after heat treatment to the crystallization temperature. A longitudinal section of the ribbon was made from the obtained sample and SEM studies were carried out. Figure 6 shows fragments of the longitudinal section of the sample from the elastic state to the plastic state.

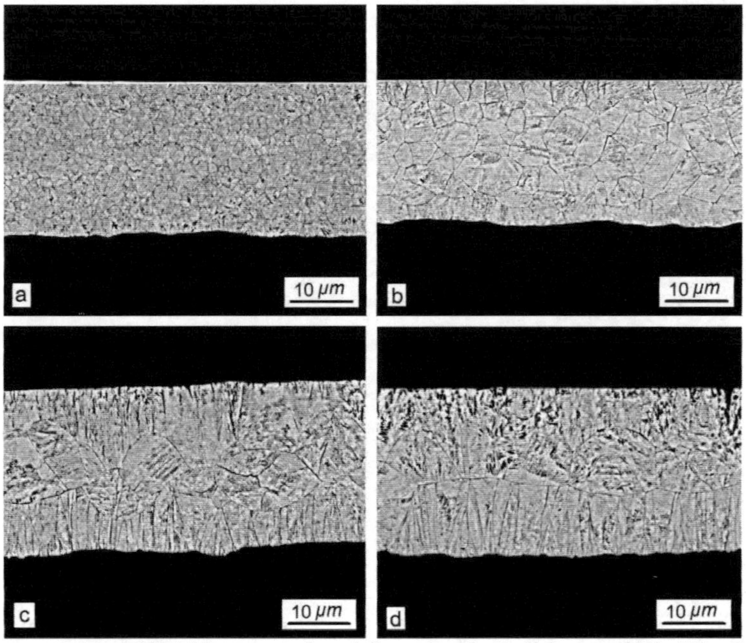

Figure 5. Cross-sectional SEM images of rapidly quenched ribbons of $Ti_{50}Ni_{25}Cu_{25}$ alloy after electropulse treatment (EPT) with 5000 ms (a), 1000 ms (b), 100 ms (c) and 10 ms (d).

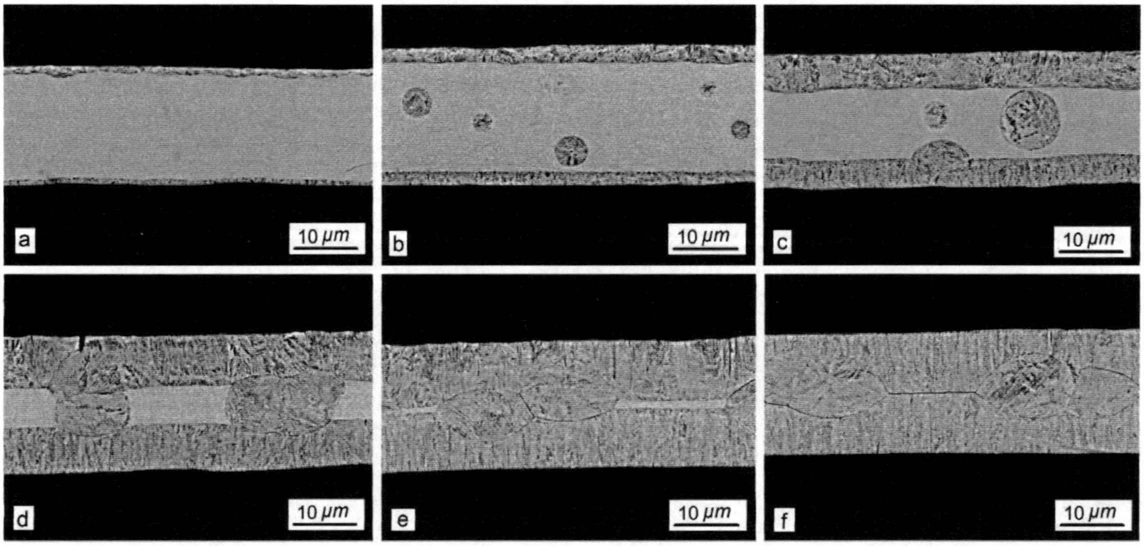

Figure 6. Fragments of longitudinal-sectional SEM images of $Ti_{50}Ni_{25}Cu_{25}$ alloy ribbons after EPT with 10 ms from elastic state (a) to plastic state (f) and intermediate parts (b-e).

It can be seen from the obtained SEM images that during the passage of an electric pulse in those areas of the sample where the current density and, accordingly, the heat energy released is insufficient for it is bulk crystallization, the formation of a crystalline phase occurs at the surfaces of the ribbon, while the amorphous phase is mainly preserved in the inner part (Fig. 6a). The formation of surface crystal layers can be caused both by edge effects that stimulate crystal formation, and by local changes in the elemental composition of the alloy at the surface of the ribbon. Where the current density is higher, surface crystals grow deep into the ribbon, and single crystalline spherulites appear in the inner amorphous part (Fig. 6b). In the regions with an even greater release of thermal energy during EPT, an increase in the thickness of surface crystals with a columnar structure is observed, and in the amorphous part, the number of nascent single spherulites and their size increases (Fig. 6c-e). Single spherulites, "clamped" in the middle part of the ribbon, take a lenticular shape. Finally, when the crystallization temperature is reached over the entire thickness of the sample, regions in a completely crystallized state are observed (Fig. 6f). Moreover, in some areas, columnar crystals from opposite sides of the ribbon directly touch in the central part of the sample. Thus, in the case of insufficient heating during EPT, different areas along the belt may have different degrees of crystallization, that is, both fully crystallized states and regions with an amorphous crystal structure are formed.

4. Conclusions

The paper shows the differences in the processes of crystal state formation in amorphous rapidly-quenched TiNiCu alloy with 25 at. % Cu by isothermal treatment, speed heat treatment and electropulse treatment (EPT).

EPT of amorphous $Ti_{50}Ni_{25}Cu_{25}$ alloy ribbons generally leads to the formation of a crystal structure with an inhomogeneous distribution of crystals over the thickness of the ribbon with surface columnar and internal bulk crystals. It is found that when an electric pulse passes through an amorphous ribbon under conditions of insufficient thermal energy density for its bulk crystallization, the formation of a crystalline phase occurs at the surfaces of the ribbon, while the amorphous phase is mainly preserved in the inner part. In this case, states with different degrees of crystallization are formed along the ribbon.

It is established that the formation of the crystal phase in amorphous ribbons under EPT occurs from the surface to the inner part due to the predominant formation and growth of surface columnar crystals and the subsequent nucleation and growth of crystals in the rest of the ribbon. Under certain EPT conditions, it is possible to form regions in the amorphous ribbon in which only columnar crystals are present, which touch in the central part of the ribbon.

Acknowledgment
The study was supported by Russian Science Foundation (grant number 19-12-00327).

References

[1] Nespoli A, Besseghini S and Pittaccio S, Villa E and Viscuso S 2010 Sensors Actuators A Phys **158** 149
[2] Morgiel J, Cesari E, Pons J, Pasko A and Dutkiewicz J 2002 *J. of Mat. Science* **37** 5319
[3] Senkovskiy B, Usachov D, Fedorov A et al. 2012 *J. of Alloys and Compounds* **537** 190
[4] Kang S, Lim Y, Lee Y, Moon H, Kim J and Nam T 2010 *Scr. Mater.* **62** 71
[5] Shelyakov A, Sitnikov N, Borodako K et al 2020 *J. of Micro-Bio Robot* **16** 43
[6] Shelyakov A, Sitnikov N et al 2018 *Acta Physica Polonica* **134(3)** 708
[7] Irzhak A, Zakharov D, Koledov V et al. 2014 *J. of Alloys and Compounds* **586** 464
[8] Sheng Y, Hua Y, Wang X et al. 2018 *Materials* **11** Art. numb. 185.
[9] Zhu R, Tang G, Shi S et al. 2013 *Appl. Phys. A-Mater.* **111** 1195
[10] Delville R, Malard B, Pilch J et al. 2010 *Acta Materialia.* **58** 4503
[11] Cao W, Zhang J, Shek C 2013 *J. of Materials Science and Tech* **29** 1135
[12] Zhu R, Liu J, Tang G et al. 2014 *J. of Alloys Compounds.* **584** 225
[13] Shelyakov A, Sitnikov N, Khabibullina I et al. 2019 *Materials Letters* **248** 48
[14] Sitnikov N, Shelyakov A, Khabibullina I et al. 2020 *Conf. Series: Mat. Sc. and Eng.* **770** 012088
[15] Shelyakov A, Sitnikov N, Menushenkov A et al. 2011 *Thin Solid Films* **519** 5314

IOP Conf. Series: Materials Science and Engineering 1213 (2022) 012010 doi:10.1088/1757-899X/1213/1/012010

Evolution of Transformation Strain During Partial Transformation Cycling of an NiTi Shape Memory Alloy

G Swaminathan[1], V Sampath[1,*]

[1]Department of Metallurgical and Materials Engineering, Indian Institute of Technology Madras, Chennai 600036, India.
*Corresponding author: vsampath@iitm.ac.in

Abstract. This paper deals with how the magnitude of transformation strain changes on partial transformation cycling of an NiTi shape memory alloy. A near-equiatomic NiTi shape memory alloy was allowed to undergo partial thermal cycling keeping the stress constant at 100 MPa for various upper cycle temperatures (between austenite start and austenite finish), using a custom-built thermomechanical cycling test setup. The displacement and the temperature of the sample during cycling were measured using a LASER extensometer and an optical pyrometer, respectively. The test results show that the recovery strain and thermal hysteresis width decrease with increasing number of cycles during partial cycling. In addition, martensite start and martensite finish temperatures increase during the initial cycles, whereas austenite start and austenite finish temperatures decrease during the initial cycles, followed by their saturation.

1. Introduction

The combination of two effects, such as shape memory effect and superelasticity, makes shape memory alloys (SMAs) attractive for engineering (robotic arms, smart actuators, etc.) and medical (orthodontic archwires and stents) applications [1,2]. SMAs are routinely used in cyclic applications, such as actuators and flaps of air conditioners, to operate repeatedly between their transformation temperatures under thermomechanical cycling conditions [3,4]. However, repeated thermomechanical cycling causes the SMAs to degrade their shape memory characteristics, such as recovery strain, transition temperatures, and thermal hysteresis. This behaviour is therefore named functional fatigue [5]. Functional fatigue is caused by dislocations that are generated by cycling (transformation-induced dislocations), and their density is usually higher at the initial cycles [6]. Generally, thermal cycling lowers transformation temperatures, whereas thermomechanical cycling raises them [7,8]. However, when the alloy is subjected to a specific number of cycles, it undergoes work-hardening by dislocation-dislocation interactions. As a result, the generation of additional dislocations becomes more difficult, resulting in stable behaviour [9]. Similarly, the recovery strain and permanent strain decreases and increases, respectively, with increasing number of cycles.

In order to improve functional fatigue behaviour, a lower hysteresis width is of prime importance in the actuator design since it enhances the speed of response [10]. In recent years, partial thermal cycling has been studied so as to enhance the shape memory properties as well as functional fatigue behaviour of SMAs. It has been proposed that partial cycling could enhance the functional fatigue behaviour of SMAs by ten times as compared to full cycling [10–13]. When the SMAs undergo full cycling, the SMAs completely transform from austenite to martensite during cooling (from above A_f to below M_f) and martensite to austenite during heating (from below M_f to above A_f). But, during partial cycling, only a fraction of the phases will undergo transformation because the stimulation (such as heat or stress) will be arrested before transformation is completed [14]. Flaps and thermal actuators work under a range of upper cycle temperatures (UCT) while in operation [15]. As a result, we expect the SMAs to perform well under the partial transformation (UCT<A_f) rather than full cycling (UCT>A_f) without any degradation in performance. Consequently, the present study is intended to explore the influence of temperature corresponding to the upper cycle on the shape memory characteristics of an Ni-Ti SMA during partial thermomechanical cycling.

2. Materials and Methods

Near-equiatomic NiTi SMA sheets of 0.5 mm thickness were used in this study. Homogenization treatment was performed at 900 °C for 3.6 ks before quenching them into a water bath at the ambient temperature. The alloy in the initial state contained coarse recrystallized grains. The transformation temperatures of the alloy were determined using a TA Instruments make Discovery 25 model differential scanning calorimeter (DSC), as shown in
Figure **1** (a). A wire-cut EDM machine was used to prepare the samples from the sheets for testing. Using a custom-built thermomechanical cycling test setup, as shown in
Figure **1**, test samples were subjected to partial thermomechanical cycling for 1000 cycles at the constant stress level of 100 MPa at different upper cycle temperatures (45, 50, 55, and 60 °C).

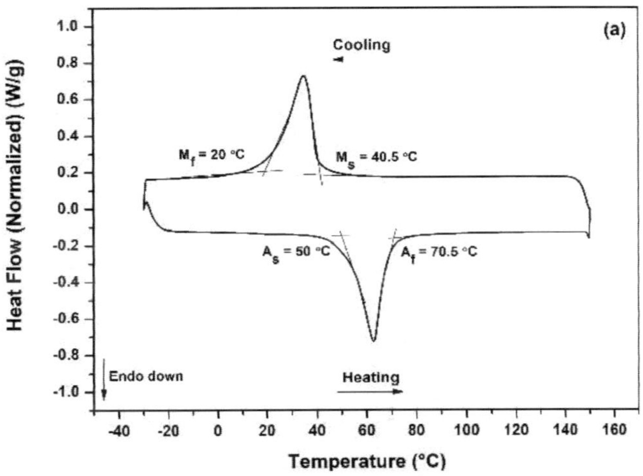

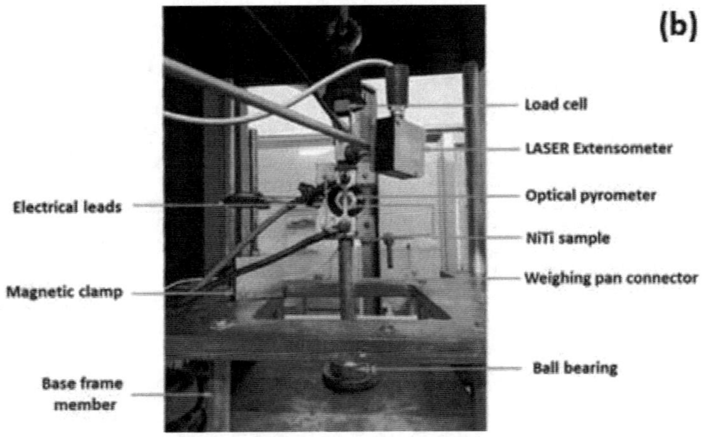

Figure 1. (a) DSC thermogram of the NiTi SMA used in the present study, (b) Photograph of the thermomechanical cycling test setup

Since the objective of the present study is to trace and understand the evolution of transformation strain during partial cycling, the above temperature range was chosen. The sample was heated to the predetermined upper cycle temperature by passing an electric current using the temperature-controlled mode of heating. Further, the sample was cooled using the cooling fans, which are attached to the frame of the machine. The testing machine was equipped with an optical pyrometer and a LASER displacement sensor that measured temperature and strain, respectively, during cycling. The data generated by the test was saved in a PC, which was connected to the machine, for analysis.

3. Results and Discussion

Based on experimental data, the transformation strain during partial cycling is calculated for different upper cycle temperatures by using the ratio of the change in length (dl) to the original length (l) and plotted against the temperature, as shown in figure 2 (a-d). The transformation strain/recovery strain rapidly decreases during the initial cycles due to the rapid increase of dislocations in the early stages of cycling, as shown in figure 3. In subsequent cycles, it gets saturated due to the difficulty of generating additional dislocations by the material due to strain hardening and displays stable behaviour [10]. In addition, the shift in the transformation strain vs. temperature plots (figure 2 (a-d)) indicates the occurrence of permanent deformation.

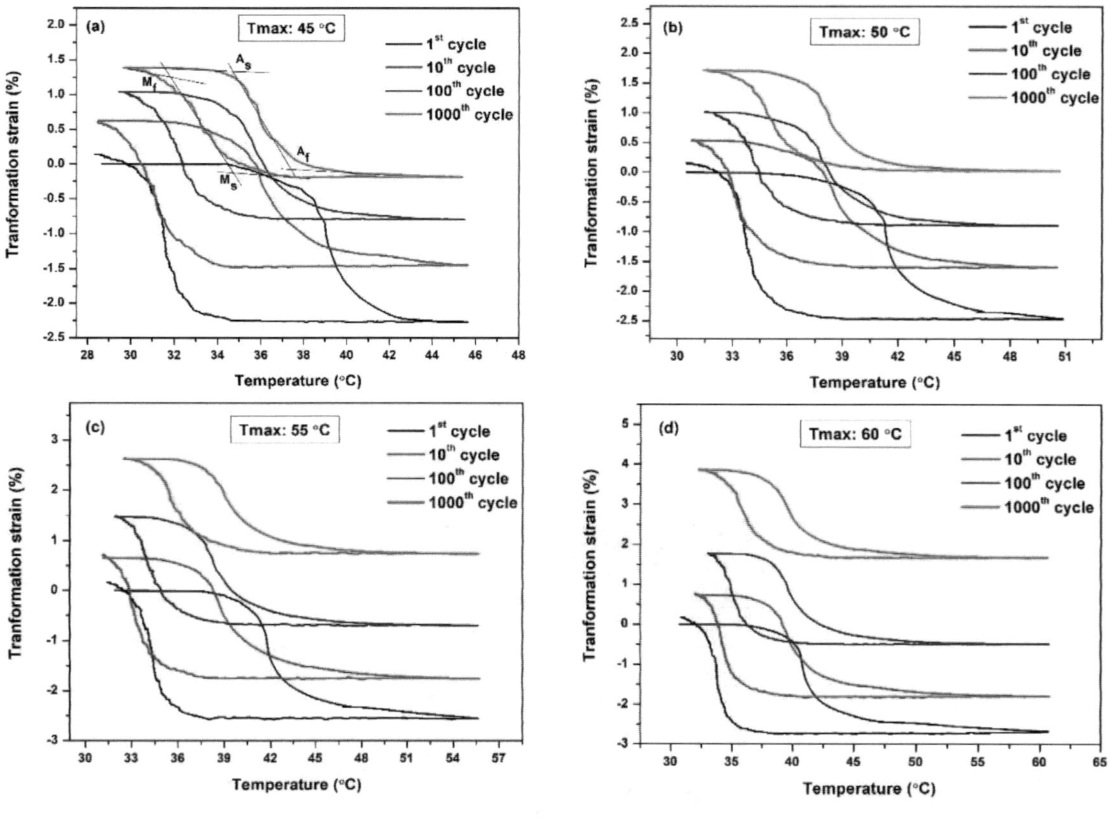

Figure 2. The evolution of transformation strain vs. temperature for different upper cycle temperatures: (a) 45 °C; (b) 50 °C; (c) 55 °C; and (d) 60 °C, for NiTi SMA under partial thermomechanical cycling.

From the strain vs. temperature curves, the characteristic transformation temperatures, such as M_s, M_f, A_s, A_f, of the alloy for different upper cycle temperatures, were measured by the tangent method, as shown in figure 2 (a), adhering to the ASTM standard E3097– 17 [16].

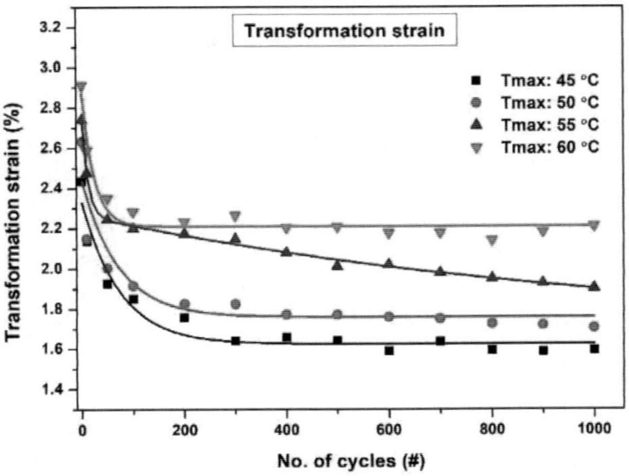

Figure 3. The variation in transformation strain with increased number of cycles for NiTi SMA during partial thermomechanical cycling.

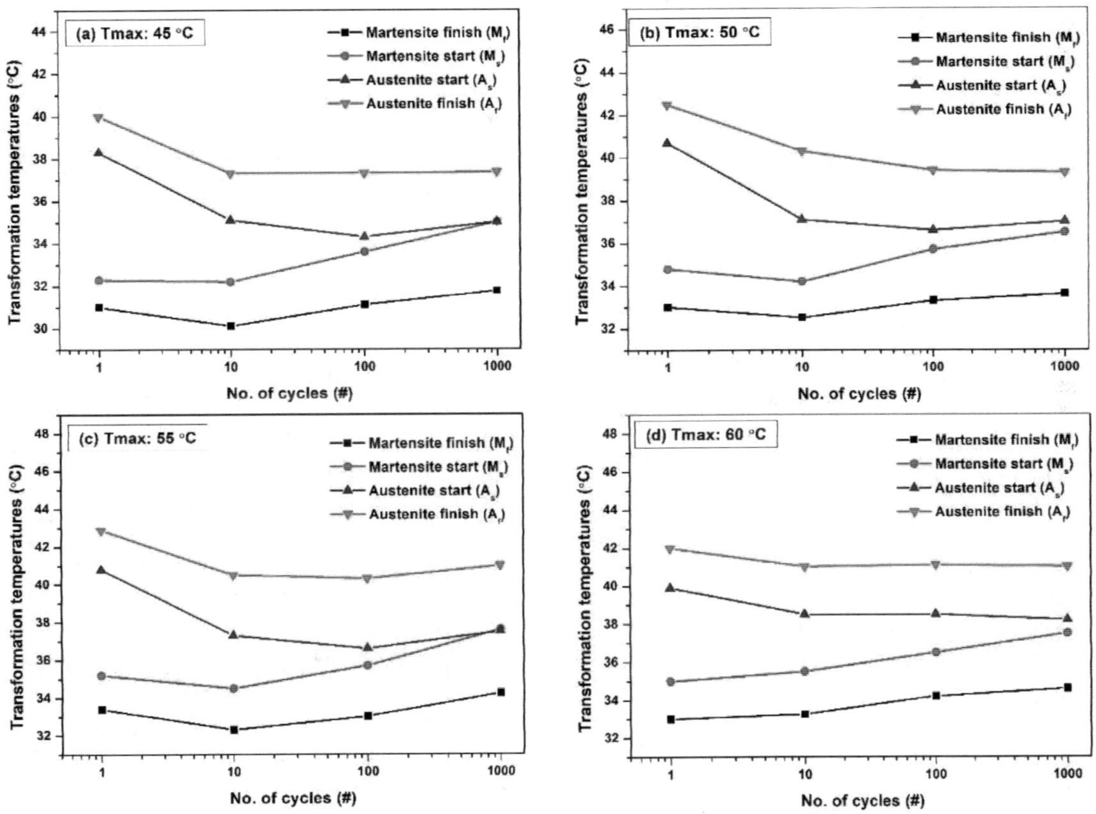

Figure 4. Effect of partial thermomechanical cycling on phase transition temperatures of NiTi SMA at different upper cycle temperatures (log scale).

From figure 4 (a-d), it can be seen that the A_s and A_f temperatures were lowered, while the M_s and M_f temperatures were increased, as the number of cycles was raised. The shift in transformation temperatures occurs due to changes in the upper cycle temperature, and these transformation temperatures correspond to the operating temperatures of the particular cycle. Especially, the decrease in the A_f temperature is attributed to the decrease in the upper cycle temperature due to the partial cycling. Moreover, the generation of dislocations during thermomechanical cycling facilitates the nucleation of the martensite while cooling, and as a result, the M_s temperature increases [17,18].

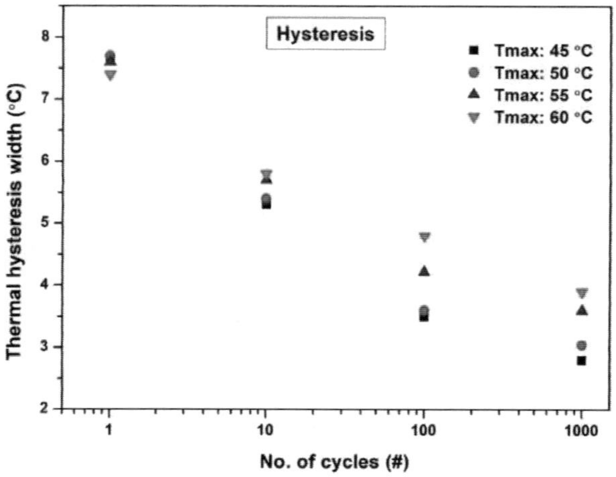

Figure 5. The influence of partial cycling on the thermal hysteresis width of NiTi SMA at different upper cycle temperatures (log scale).

In all cases, it is observed that the transformation temperature window (M_s to A_f) decreases as the number of cycles increases. Additionally, M_s and A_s temperatures tend to converge as the cycle number increases, as shown in figure 4 (a-d). A major finding of this work is that the reduction in the transformation temperature window may enable SMA-based smart actuators to achieve higher actuation rates. Additionally, the reduction in the transformation temperature window leads to the reduction of thermal hysteresis width during cycling.

From figure 5, it is observed that the width of the temperature hysteresis is lowered for all cases as the number of cycles increased. Nevertheless, the reduction is higher for the lowest upper cycle temperature and lower for the highest upper cycle temperature. Generally, an increase in the functional fatigue characteristics of an SMA is associated with a decrease in hysteresis width [10,19]. This result, therefore, suggests that the narrow transformation temperature window achieved by the partial thermomechanical cycling can improve the functional fatigue behaviour of NiTi SMA with an improved fatigue life.

4. Conclusions

The near-equiatomic NiTi SMA was subjected to partial cycling at different upper cycle temperatures under constant stress. The following conclusions have been drawn from this study.

1. The transformation temperatures A_s and A_f are lowered, whereas the M_s and M_f are increased as the number of cycles is raised. It is attributed to the change in the upper cycle temperature and number of dislocations, which serve as potential nucleation sites for martensitic transition.
2. The recovery strain is lowered as the number of cycles is raised for all cases, and the rate of reduction is higher in the initial cycles due to the generation of dislocations, as their numbers tend to stabilize after a specific number of cycles.
3. Thermal hysteresis width during the forward/reverse transitions decreases as the number of cycles is increased, and the lowest hysteresis width corresponds to the lowest upper cycle temperature.
4. This study shows that partial thermomechanical cycling is one of the potential strategies to increase the actuation rate of an SMA-based smart actuator with an improved lifespan.

Acknowledgement

The authors would like to acknowledge the Science and Engineering Research Board, Department of Science and Technology India, for the financial support under the grant number CRG/2019/002267.

References

[1] Duerig, T W, Melton K N, Stockel D W C M 2013 *Engineering Aspects of Shape Memory Alloys* (Butterworth-Heinemann Ltd.)
[2] Humbeeck J Van 1999 *Mater. Sci. Eng. A* **273–275** 134–48
[3] Otsuka K, Saxena A, Deng J and Ren X 2011 *Philos. Mag.* **91** 4514–35
[4] Pelton A R, Schroeder V, Mitchell M R, Gong X Y, Barney M and Robertson S W 2008 *J. Mech. Behav. Biomed. Mater.* **1** 153–64
[5] Eggeler G, Hornbogen E, Yawny A, Heckmann A and Wagner M 2004 *Mater. Sci. Eng. A* **378** 24–33
[6] Van Humbeeck J 1991 *Le J. Phys. IV* **01** C4-189-C4-197
[7] Miyazaki S, Igo Y and Otsuka K 1986 *Acta Metall.* **34** 2045–51
[8] Pelton A R, Huang G H, Moine P and Sinclair R 2012 *Mater. Sci. Eng. A* **532** 130–8
[9] Richards A W, Lebensohn R A and Bhattacharya K 2013 *Acta Mater.* **61** 4384–97
[10] Lagoudas D C, Miller D A, Rong L and Kumar P K 2009 *Smart Mater. Struct.* **18** 1–12
[11] Liu Y, Laeng J, Chin T V and Nam T H 2006 *Mater. Sci. Eng. A* **435** 251–7
[12] Tang W and Sandström R 1993 *Mater. Des.* **14** 103–13
[13] Liu Y, Laeng J, Chin T V. and Nam T hyun 2008 *J. Alloys Compd.* **449** 144–7
[14] Liu T, Zheng Y and Cui L 2013 *Mater. Lett.* **112** 121–3
[15] Strelec J K, Lagoudas D C, Khan M A and Yen J 2003 *J. Intell. Mater. Syst. Struct.* **14** 257–73
[16] ASTM International 2017 *ASTM Stand.* E3097-17
[17] Fukuda T, Saburi T, Doi K and Nenno S 1992 *Mater. Trans. JIM* **33** 271–7
[18] Ibarra A, San Juan J, Bocanegra E H and Nó M L 2007 *Acta Mater.* **55** 4789–98
[19] Huang W 2002 *Mater. Des.* **23** 11–9

AUTHOR INDEX